Geomatics for Cultural and Natural Heritage Conservation and Valorisation

Agata Lo Tauro

BAR International Series 2711
2015

Published in 2016 by
BAR Publishing, Oxford

BAR International Series 2711

Geomatics for Cultural and Natural Heritage Conservation and Valorisation

ISBN 978 1 4073 1362 7

BAR Publishing is the trading name of British Archaeological Reports (Oxford) Ltd. British Archaeological Reports was first incorporated in 1974 to publish the BAR Series, International and British. In 1992 Hadrian Books Ltd became part of the BAR group. This volume was originally published by Archaeopress in conjunction with British Archaeological Reports (Oxford) Ltd / Hadrian Books Ltd, the Series principal publisher, in 2015. This present volume is published by BAR Publishing, 2016.

Printed in England

BAR
PUBLISHING

BAR titles are available from:

BAR Publishing
122 Banbury Rd, Oxford, OX2 7BP, UK
EMAIL info@barpublishing.com
PHONE +44 (0)1865 310431
FAX +44 (0)1865 316916
www.barpublishing.com

Contents

Please note that the figures are available to download from www.barpublishing.com/additional-downloads.html

ABSTRACT

This project is one of the first systematic attempts in developing a Geographic Information System (GIS) for Cultural landscape and heritage with the support of several geomatic applications. The Etna Park has been chosen for the case-study but the framework established allows easy extension to other parks.

The first objective of this study was to investigate data availability for GIS applications in the Etna Park. It was found that data of the green infrastructure and cultural heritage available on the Etna Park, are useful for the GIS application and implementation of cultural itineraries. In this context, a basic requirement for the facilitation of the GIS is the ability to be able to detect and extract indicators of natural and anthropized territories, in particular the greenway network and the cultural landscape and heritage including areas of D.O.P and IGP production, protected vegetation, green monuments, agricultural parks, protected areas (such as SIC and ZPS), from available satellite remote sensing (RS) data. Subsequently, the indictors of growth derived from RS imagery can be integrated into a multi-source GIS database with ground-based data and land use analysis to facilitate potential environmental management analysis. The extraction of useful data from RS imagery for GIS-based analysis of the Etna park is achieved through an integrated conceptual and operational framework presented in the book. Land use coverage from satellite imagery was useful for its spatial resolution and, in particular, for the semi-automatic approach for the classification of greenways. Major investment is required to generate and analyse the land use by digitizing aerial photos for change detection analysis. A few significant data limitations have been identified (small hiking trails, water lines, public arts, cultural heritage, outdoor facilities, etc.) and should be considered in the future development effort. The second objective of this project was to develop a methodology within the PC windows environment to facilitate the analysis and application for the park staff who are not familiar with the GIS software, GISCloud and to implement educational activities in this field of study. The custom-design has been made with the ENVI and ESRI technologies. The views generated by the system can be printed as hard copy maps. The program structure can be used for other park's databases in Europe. The final objective was to demonstrate how the GIS (including GISCloud) may be used in helping to address multi-objective land use and planning issues facing the Park Departments in Europe. Thematic maps have been created using buffer and map overlay analysis to guide the zoning and planning efforts for protection of sensitive cultural and natural resources.

ACKNOWLEDGMENTS

There are many individuals and organizations that deserve thanks for having provided contributions and support in the development of this research. I would also like to thank Ing. G. Di Paola and S. Spina for their useful inputs in terms of imagery data, edits and suggestions that partially helped polish the final document into mapping GIS implementation. Thank you also to my other readers, colleagues for their excellent and positive advise as I worked through my studies (including Catania and Trieste University).

A large thank you goes to the Etna park – the GIS department – the National Department of Civil Protection, Department of Civil Protection (Nicolosi) and the Catania Province staff for assisting me with the support that facilitated my field research in the GIS and RS. Without this opportunity, my work in the Etna Park would not have been possible.

A special thanks goes to the Etna Park Board of Directors, and the staff at the INGV Department (Centre for Research on Remote Sensing at Rome), for their availability during my study. Also, many thanks to the staff at the Catania Province for providing me with sources, imagery, data and sometimes a place to work while at Catania Province.

I would also like to offer thanks to Fabrizia Buongiorno (Researcher at the INGV, Rome) who was available to share information. I would also like to thank Daniele Bartolucci (Geotop Positioning System), Michele Gagliano (C.G.T srl), Massimo Zotti (Planetek) and other Institutions for their useful input in terms of open sources, imagery data and survey that helped polish the final realization of 3D - 2D representation and visualization. It is to all my family and to my father ("fu" Dott. Ing. Antonino Lo Tauro) that all my research is dedicated.

CHAPTER 1
INTRODUCTION

Documenting cultural landscapes or landscapes around objects of cultural heritage *has often been* studied with some practical or theoretical aim in order to find novel stategies in the field of Geomatics (Lo Tauro, A. 2010). In a systematic way the complex topics and techniques are covered that can be assembled under geomatics namely, geodesy, cartography, photogrammetry, remote sensing (RS), informatics, acquisition systems, global positioning systems (GPS), digital image processing (DIP), geographic information systems (GIS), decision support systems, and web GIS. Others include: Telecommunications, Robotics, Geo-spatial Information, Information and Communication Technologies (ICT).

With modern surveying and visualization techniques, landscape documentation, and visualization has become affordable and rewarding. Especially when remote sensing is combined with geographic information systems, virtual landscapes can be a powerful means to describe a landscape and to monitor and manage its conservation and development.

The need for access to accurate, up-to-date land use information is especially important for urban land use management in parks. To help develop policies targeted at ameliorating these problems and facilitating environmentally sound development, planners are increasingly turning to the use of spatial information technologies such as remote sensing (RS) and Geographic Information Systems (GIS). However, there is a dislocation between policy development and planning, relative to the potential contribution offered by spatial information technologies. This is especially true in the realm of sustainable environmental management (SEM) and cultural heritage conservation. This study explores some of the needs of SEM through the use of spatial information technologies. The tools of RS and GIS are combined in a temporal analysis to demonstrate their use for the identification of potential environmental management indicators in targeting problem areas on the Etna Park in Sicily.

Planners must deal with this problems using analytical approaches suited to set indicators for an environmental planning. The core of such analytical approaches is the integration of high spectral and spatial resolution remote sensed data with other ancillary data in GIS environment. The GIS will be implemented as a basis for the spatial representation of environmental and historical-cultural heritage in the area of Etna volcano, Sicily. This study provides an analysis of potential ecological and cultural connectivity to identify and link areas with priority conservation significance and landscape linkages as part of the Etna Park program called the greenway network.

This study also presents an application of MIVIS airborne spectral scanner data and remote sensing data. The research implemented an "image based" classification approach of MIVIS data and other sources to recognize green areas (trees, grass, etc.) and impermeabilizated areas (cultural heritage, traditional building, some routes and greenways) using ENVI analytic software. The next step was the integration of MIVIS layers with land use and land plan GIS data into GIS environment and GISCloud to produce, using topological overlay functions and GIS analysis, a set of indicators that point out ecological target values for an environmental planning.

1.1 Problem Statement

The rapid and often disorganised sprawl of parks causes many difficulties for their inhabitants and makes the task of providing urban infrastructure and services a major challenge for planners for the implementation of cultural tourism. Traditionally, the means for determining the nature of new conservation and valorization strategies has been through extensive *in-situ* fieldwork. Following this, new developments are planned and parks spread outwards incrementally from their initial development. However, ground-based fieldwork is both time consuming and of questionable benefit, as the rate of change is often rapid and adherence to formal urban planning principles in infrastructure and service provision is, at best, *ad hoc*.

To facilitate properly managed development, many types of up-to-date information are required. One widely used method of obtaining timely information about the morphology of rapidly changing environmental areas is through the use of space-borne satellite imagery. High-resolution satellite imagery has the ability to provide relatively inexpensive land-cover data for mapping, inventory, and change detection at short time intervals. Satellite images provide a good overall 'picture' of an environmental area and provide a great deal of spectral information regarding land-use over extended areas. However, satellite imagery does have limitations and it is not, in and of itself, sufficient to derive useful information concerning the morphology and land use structure of regional parks. Accordingly, one of the main purposes of this research is to demonstrate the ability to transform this imagery into usable data within a multipurpose GIS environment. Once transformed into this format and combined with other ground-based information, the combined data are used to identify 'triggers' that characterize the location of zones of potential environmental values and ecological target values that require management strategies to cope with subsequent cultural heritage strategies. Temporal analysis of the listed areas in search of such zones is integral to this study.

1.2 Objectives

The analysis of land-use morphology through the integration of RS data with ground-based data stored in a GIS offers improved insights into both land structure and the process of land-use change.

This reasearch focuses on this integration to facilitate the study of land use change signals, especially at the listed areas (such as SIC and ZPS, under the UE Project "NATURA 2000 – Direttiva Habitat 92/43/CEE). The general objective is to develop an operational methodology for use by planners and others employing the tools of RS and GIS to identify areas of potential environmental values and ecological target values associated with the conservation and valorization strategies for the cultural heritage. The specific objectives of the research are:

(i) to develop an integrated conceptual framework based on the concepts of cultural tourism and sustainable environmental management for the achievement of SEM in the Etna park;

(ii) to operationalise this conceptual framework through the integration of RS and GIS spatial information technologies,

(iii) to extract greenways and classify adjacent land uses from temporal RS imagery and, subsequently, to put these features into a GIS for further temporal analysis with ground-based data to define the greenway network;

(iv) to detect land use change and subsequent indicators of potential environmental values and ecological target values;

(v) to generate a geo-database in order to take measures to facilitate strategies for assistance intervention and other intervention support for the listing and conservation of cultural heritage;

The case study for the research is the Etna Park, Sicily. In this context, indicators of stress and ecological target values, which are identifiable using both RS and GIS technologies, are used to control and limitate the process of environmental management and potential locations for the incidence of environmental stresses and stimulate the e.tourism and cultural tourism.

Specifically in this research the green infrastructure construction is regarded as the initial trigger for generalised land use change around which growth follows. The main notion here is that any sprawl must initially begin with some form of path or greenway to allow for greater access and mobility. Consequently, where greenway network develop, people soon follow as areas that were formally inaccessible to vehicular traffic are opened up. Increased population then leads to the need for improved infrastructure provision and other goods (historical, cultural and natural resources) and services (connection nodes such as stations, parking or restaurants, hotels and assistance points). In this sense, greenways are identified as the initial indicator of environmental sources that 'trigger' further development.

To achieve the overall objectives of the research, a procedure is developed for extracting trigger-related features from imagery and converting these features to data structures that can be used with a GIS for urban environmental management. Emphasis is placed initially on greenways networks, followed by the incorporation of other indicators within a multipurpose GIS. Because this analysis is completed at two different points in time, the second specific aspect of the contribution made by this book is to provide a the definition of methodologies for the analysis of cultural heritage, important cultural sources for the territory of the Etna Park. For this purpose it was implemented a project GIS specifically oriented to the conservation and valorisation of cultural heritage (L. 137/02) and to the implementation of Informative Systems for larger regional programs. The result of this research is the realization of a database for specialised GIS platform in order to manage, in real time, those sources and to develop novel methodologies for the listing and georeferencing of cultural heritage. This tool for the management of different sources and for the updating and the implementation of the database can be useful for the Etna Park Institution or other Public Institutions. Those kind of analysis provides important suggestions.

1.3 Structure of the book

In Chapter 2, literature on RS and GIS for the environmental management, cultural tourism, cultural heritage conservation and their intersections is reviewed. This discussion comprises multiple aspects of the concept and principles of sustainable environmental management (SEM). The review further considers the role of spatial information technologies in fostering environmental planning in parks. The literature review combines the use of satellite remote sensing and GIS information technologies to make operational the conceptual model of SEM and Cultural Tourism (CT). Chapter 3 builds from Chapter 2 by presenting a detailed description of the research. The study areas selected in the Etna Park are discussed and the methods of analysis used are summarized.

Chapter 4 presents and discusses the results. The accuracy of the remote sensing (RS) imagery analysis, including greenway feature extraction and temporal land use change analysis, is assessed and the subsequent integration of the extracted image features is analyzed with cultural heritage resources and data. This allows assessment of the effectiveness of the combination of RS data , GIS and GISCloud to detect land use change and subsequent indictors of potential environmental stresses and cultural values.

The methodology establishes a prior semi-automatic approach for the research on vegetation which is suggested to become an ecological corridor (usually a narrow structure of vegetation).

Chapter 5 presents the conclusions derived from the results of the research. Some recommendations for further research are presented pertaining to the methods and the overall effectiveness of the process.

CHAPTER 2
SUSTAINABLE ENVIRONMENTAL MANAGEMENT & SPATIAL INFORMATION TECHNOLOGY

This chapter first discusses the concept of sustainable development. The role of sustainable development in policy and planning at various levels is then discussed, with a focus on how the concept is applied in environmental planning. Next the concept of environmental management is explored. Subsequently, both concepts are combined into a unitary entity termed sustainable environmental management (SEM). This new concept is then developed in the context of modern information technologies, especially those that focus on visualizing and analyzing spatial data relative to environmental management E.Tourism and Cultural Tourism (CT).

The following is not a list of articles found in a variety of publications but a literature review. I have tried to provide a summary for each of the articles and have noted the journals where these articles could be found. The integration of Remote Sensing (RS) and Geographic Information Systems (GIS) is first reviewed, with subsequent individual focus for each in terms of their use. In particular, the use of RS for skeletal greenway extraction is reviewed in detail. From these two data sources, the generation of a multi-purpose, and novel multi-source GIS databases discussed.

To unite the main components of the preceding review, an integrated model for the role of the GIS in the case-study is presented. This overall model demonstrates the linkages between SEM, CT and SIT by building on the concepts of sustainable development, urban environmental management E.Tourism and Cultural Tourism and defining the lineage between these concepts and the use of SIT to operationalize the process of environmental stress detection and valorization of cultural heritage. Lastly, based on the operational portion of this model, a related operational model for the use of SIT in implementing the concepts of SEM and CT is presented and discussed. This section also underlines how the introduction of digital information technologies such as novel Geographic Information Systems (GIS) influences both the preservation and valorisation of cultural heritage and attraction consumption. The use of potential indicators of environmental stress and for tourist ecological target values derived from RS and ground-based GIS data are then reviewed in accordance with their use in this model.

2.1 Sustainable Development (SD)

The term development started to appear in common parlance shortly after the Second World War. The concept of SD was introduced by the International Union for the Conservation of Nature in 1980 when they revealed their framework for world conservation (Portnov and Pearlmutter, 1999). SD can be defined as a shift from singular economic policies designed simply for maximizing economic growth to policies designed to reflect the need to consider 'sustaining' natural resources and taking into account social and cumulative impacts of development (Tjatera, 1994). The United Nations (1996) defines SD as a process that meets the needs of the present without compromising the ability of future generations to meet their own needs. Based on both Tjatera and the United Nations (UN) definitions, it is easy to see the need for such policies in guiding SD. The publication

'Our Common Future', by The World Commission on Environment and Development in 1987, has become the definitive statement on global sustainable development and the environment. The commission recommends seven strategic imperatives designed to allow nations to move away from destructive processes of growth and development and to follow a path of SD that allows environmental policies and development strategies to be integrated. These include:

1) reviving growth;

2) changing the quality of growth;

4) ensuring a sustainable level of population;

5) conserving and enhancing the resource base;

6) reorienting technology and managing risk;

7) merging environment and economies in decision making.

(WCED, 1987)

Ultimately, because SD often only occurs with painful opportunity costs its procurement must rest on the political will of governing bodies. In the broadest sense, the strategy for SD endorsed by the World Commission sought to promote harmony between humanity and nature. In the pursuit of SD, and to meet the strategic imperatives noted above, they specifically noted the need for improved technological capabilities. The use of IT (specifically SIT) in DC is addressed in a subsequent section of this chapter.

The concept of SD can be divided into its two parts, sustainability and development.

Portnov and Pearlmutter (1999) further subdivide sustainability into literal, ecological, and social aspects. These aspects respectively represent the sustainability of anything that comprises the ecological basis of human life and the social basis of human life respectively.

In the same manner, they sub-divide development into process, growth and change and objectives. Process and growth and change can be interpreted as sustaining growth while objectives represent the attainment of basic needs, both in terms of the semantics of sustainable development. The concepts of SD addressed here provide the framework necessary for practising sustainable urban development. These include environmental, economic and socio-demographic dimensions. Each of these dimensions must be addressed for SD to occur in natural reserves, parks and environmental context.

Considerable progress has been made in the last ten years on innovation and implementation of policy in regard to the goals of SD. The ability to assess countries in terms of their sustainability and their adherence to growth goals that subscribe to the principles of sustainability is difficult. To assist with this, Satterthwaite (2001) outlines five broad categories of environmental action within which the performance of all countries should be assessed. These include:

1) achieving a high quality environment for all inhabitants (e.g. park space);

2) minimizing the transfer of environmental costs to the inhabitants and ecosystems;

FIGURE 2-1-1: RS FOR SD DEVELOPMENT AND FOR THE IDENTIFICATION OF AREAS OF POTENTIAL ENVIRONMENTAL VALUES AND STRESS.

3) ensuring process towards sustainable consumption (this can be defined as ensuring people's consumption needs without undermining environmental capital).

The first three points fall under the realm of meeting residential needs, while the latter two are more problematic in that they have environmental impacts and are difficult to administer through the conventional mandate of local authorities (Satterthwaite, 2001).

These five indicators, while useful, are still difficult to use. The primary reason for this lies with the unavailability of data of sufficient quality required to undertake the type analyses required. For SD development to occur, areas of potential environmental values and stress that may emerge need to be detected quickly and accurately analysed (figure 2-1-1).

The various disastrous human interventions, such as flow variations, excavation works and dike building and the subsequent modification that occurs in terms of new and existing infrastructure to support conservation strategies, is the driving force behind the creation of these areas of potential environmental values in order to limit the effect of environmental stress.

Hence, SD is needed to ensure that proper planning and development policies can be applied to these areas to promote orderly development and prevent environmental degradation.

SD has become a major priority in the public and private sectors in recent years in order to maintain social quality of life and support economic development while preserving the environment. In order to support SD in urban areas, local governments adopt various strategies such as land use planning that require the quantification of SD performance to ensure the success of such these strategies. Although several proposed a number of indicators for evaluating SD in urban areas, there is still a need for a set of metrics and a model for decision makers to make a robust quantification of SD. The main objectives of this study are to develop: (1) a set of metrics and a quantitative model for evaluating SD in urban neighborhoods efficiently; and (2) a prototype decision support tool to facilitate the application of the developed model by decision-makers (Karatas, A. and El-Rayes, K., 2012)

An other important contribution to the sustainable development programme is the Global Monitoring for Environment and Security (GMES http://www.gmes.info/).

The GMES initiative was launched in May 1998 in Baveno and adopted by the ESA and EU Councils respectively in June and November 2001. Member States have shown a keen interest in GMES from the early days, by hosting several high-level conferences and workshops to further progress on the GMES definition (Lille, Stockholm, Brussels, Noordwijk, Athens and most recently Baveno).

The overall aim of the GMES initiative is to support Europe's goals regarding sustainable development and global governance by providing timely and quality data, information, and knowledge.

The results and recommendations of the GMES Initial Period (2001-2003) are the fruit of a close and sustained joint effort by the European Commission and the European Space Agency Executive. This includes collaboration within the GMES Support Team to ensure the complementarity of research activities within the 6th Framework Programme for Research and the ESA GMES Service Elements activities and mutual assistance in the activities related to the GMES Steering Committee and GMES Forum. Continuing collaboration is planned throughout the proposed GMES Implementation period (2008-2012), taking appropriate advantage of the ESA / EC Framework Agreement. Access to information has strategic value in the development of nations and regions. This entails the capacity to have independent access to reliable and timely information on the status and evolution of the Earth's environment at all scales, from global to regional and local. GMES must also ensure long-term, continuous monitoring on a time-scale of at least decades.

In particular the information delivered by GMES will serve the three main themes:

- Meeting Europe's environmental obligations: contributing to the definition, negotiation, implementation and verification of the European environmental policies, national regulations and international conventions;
- Supporting sustainable development both within EU territory and globally. This requires the integration of the environmental dimension in European policies from other Directorate-General's of the Commission, such as agriculture, forestry and fisheries, transport, regional development, external relations including humanitarian and development aid, and possible other policies at national and international level;

- Contributing to the security of citizens both within Europe and globally, by providing the information support to decision-makers and to operational actors; Europe's ambitious global environmental monitoring programme will now be able to move into operation thanks to the European Parliament's approval of the multiannual financial framework budget for 2014–20.

2.2 Role of Spatial Information Technology in Facilitating the Goals of SEM

Urban development and land management are also processes that, if implemented properly, can benefit from the use of RS and GIS technologies. Specifically, Bishop (2000) and Abe A. (2012) point to the success of small strategic applications that can help to establish the value of GIS. Small, targeted applications are often more successful than a complex and comprehensive GIS due to the lack of well trained and educated persons in planning roles. With targeted applications, such as the extraction of green infrastructure features from RS imagery and ancillary data analysis for growth detection and potential urban stress identification, the process and results are produced more quickly than an encompassing land information system designed to incorporate too many items at too many levels of analysis.

As noted by Elkington and Shopley (2000) and Shirowzhan, S. and Lim, S. (2012) information is the ultimate renewable and therefore sustainable resource. However, as already observed, without adequate and accurate information, urban planners have difficultly making important decisions. In terms of planning-related issues, spatial information is fundamental to accurate decision-making.

As noted in the previous section, IT, SIT, and their implementation feed the process of SEM while, in turn, SEM contributes to the processes involved in implementing SIT. This process is now addressed.

2.2.1 Information Technology for SD and SEM

In recent years, advancements in microelectronics-based information and communication technologies have contributed to economic growth (UNCSTD, 2002). However, many gaps still exist. Agenda 21 (United Nations, 1992) outlines the gap between available data quantity and quality in regards to the desire to achieve more cost effective and relevant data collection. Further, Agenda 21 notes the need to strengthen the capacity of data collection and decision-making, to develop a means of ensuring planning for sustainable development, and finally, to make relevant information accessible. In general, these are desirable, yet very broad goals. While considerable amounts of data do exist, the majority of these data are not useful in many cases due to uncertainties related to data validity and accuracy.

Five activities were identified in Agenda 21 to assist in the achievement of the above goals. These activities include the development of a set of suitable SD indicators; the improvement of data collection and use; the improvement of data assessment and analysis techniques; the development of an information framework, and the encouragement of traditional knowledge wherever possible (UNCED, 1999). Breheny and Rookwood (2003) and Ruza, J., et al. (2012) list key areas that form the basis for developing a strategy to ensure future urban sustainability, namely natural resources, land use and transport, energy, and pollution and waste management. However, relatively little is being done in terms of using these as indicators to monitor SD.

Additionally, Agenda 21 notes the need for improved data collection to facilitate more robust methods of data assessment and analysis. One method of data collection that has been employed for several decades is satellite remote sensing (RS). However, its effectiveness in environmental setting related to urban areas has been limited in the past by the low spatial resolution imagery. As spatial resolution increases, the morphological structure of cities in terms of density, textural form, and spread become more visible and hence, theoretically, more easily detectible and extractible from satellite imagery. In terms of data analysis, Agenda 21 recommends new SIT to fulfil the needs of analysis of data from satellite sources. A review of the literature pertaining to existing use of RS in an environmental setting is addressed in a subsequent section.

2.2.2 Spatial Information Technology Penetration

IT, and particularly SIT, is readily utilized in different contests. However, the *effective* implementation and use of SIT is not nearly as extensive. Many pre-and post conditions must be in place before SIT penetration can be effective. For SIT to penetrate into economies, it first needs to be able to make use of local materials and persons whenever possible. This is often difficult with GIS, as training is required to make users proficient with the software. Ramasubramanian (2003), Burrough (2002), note a lack of trained personnel as one of the key barriers to GIS implementation in public institutions.

In some cases, the symbolic value of owning a GIS and its associated social status also plays a role in effective implementation and use (de Man, 2000). This can have two effects. First, the social status may be more important than the actual use and results achieved from the GIS. de Man (2000) points out that this and other non-concrete results from GIS implementation can have a significant and positive effect on organizations.

Second, the associated status can also be used as motivation to derive results from the GIS implementation.

It should be noted that Hall (2003) qualifies the above statements by noting the fact that these principles of appropriate technology focus mainly on production technology. It is possible that the differences in production technology and SIT may be too great to allow for a direct comparison in this manner. The process of exstending this analysis to the coastal zone has beguin with some early papers presented at ESA cultural events and International magazines (Lo Tauro, 2010) or other analysis (e.g. http://www.coastalt.eu) around the turn of the century, when several researchers around the world were starting to come up with new ideas and to explore some of the techniques necessary to make it work. The major space agencies and funding agencies have recognized the importance of the topic and are supporting it via a series of projects (e.g. COSTAL, PISTACH, RECOSETO and some OSTST projects)

Other factors including initial expenses, capital costs, and import duties may also render the comparison somewhat fallacious.

2.2.3 Local SIT Development for SEM

According to the American Demographic Centre (CELADE), there are six key functions that technology must facilitate when employed at the local level. These include:

1. processing of land use analysis and microdata for user-defined small-areas;

2. multi-sector databases;

3. utilization of present with past data;

4. digitized cartography;

5. spatial display and analysis of the data on maps; and

6. small-area population estimations and projections. (CELADE, 2000)

However, several barriers persist in realising these functions, especially in terms of data collection processes and data validity.

The United Nations (1996) noted that for the process of development to occur in a sustainable manner, data integrity is crucial, there is a need to improve on the lack of quality data for decision support analysis. Inconsistencies in much of the data currently available and a virtual absence of any metadata only compound the inability to determine the validity and/or accurateness of these data.

In addition to problems with data quality and availability is the fact that GIS technology may not fully meet the information needs of planning or SEM. Couclelis (2001) lists four basic functions in this regard, namely operational, management, strategic, and communication and she states that it is possible that the concept of geographic space that underlies underlying GIS technology may be quite different to the notion of space embodied in managing environmental areas. She notes that planning's notion of space can include spatial and non-spatial frameworks, such as legal and fiscal matters, as well as quantitative (land use data) and qualitative (verbal descriptions) spatial information. Within this context, important differences between site and situation are addressed. While site represents the attributes of a specific location, situation "is what characterizes a location by virtue of its being part of a spatial structure involving other locations" (Couclelis, 2001). Current SIT can address site effectively using its spatial and location linkages (GPS, co-ordinate systems, etc.). However, it is limited in its ability to provide the descriptive characteristics of a given situation. Until situational factors, such as neighbourhood type and local perspectives can be included in the decision making process, no level of site analysis can be entirely accurate or conclusive.

To address these points in more detail, the past and present operational use of SIT needs to be thoroughly reviewed and implemented (Abe, A. 2012). As such, the next four sections review the operational use of SIT to achieve SEM from RS, GIS and integrated analysis perspectives.

2.2.4 Operationalizing SEM with SIT

As noted in the previous section, the process of operationalizing SEM continues to be fraught with difficulties in many countries. However, it is argued here that, when used appropriately, IT and SIT tools can make an important contribution in making the concepts of SD and SEM operational. In this context, one of the main objectives of this research is to create a reliable process of greenway network identification, extraction and subsequent integration within a multi-source and multipurpose GIS database to help facilitate SEM. There is, however, quite widespread use of SIT to address related issues and the following sections review the use of existing RS and GIS applications and their findings. The purpose of this is to show the lack of an adequate process of integrating RS imagery and ground-based GIS data to generate indicators of areas of potential environmental values and stress.

The inclusions of RS and GIS for SEM, flow from the need to utilize SIT to achieve SEM. Proposed temporal RS analysis steps in the detection of land use change and subsequent integration with continued GIS analyses. Through closer examination of existing RS and GIS applications in the next three sections, evidence of the need for improved processes of analysis and subsequent integration of the several technologies is further revealed (Lo Tauro, A. 2012).

Given the wide productive use of SIT, it is now important to understand how SIT has been used in the past to facilitate better development policies and planning in terms not only of SEM but also by Cultural Tourism for the conservation and valorization of and environmental sources and cultural heritage in general.

2.3 E.Tourism, Cultural Tourism and New technologies

- It is vital that the Tourism Sector in emergeing markets be aware of the changes to the way the e.tourism is being implemented around the world and the implications this has for the way cultural heritage is valorized. After a decade of gathering information online - we are now beginning to actually use it in our every day lives - as less of a tool and more of a lifestyle accessory. Web 2.0 is the participatory internet. This revolution was created by the programming language XML (extensible markup language) which has changed the way the web works. This new way of coding the web has meant that content can now be freed from the restraints of form and structure, allowing content to be shared, published and distributed without its surrounding web page format with mobile devices.

  In a Web 2.0 environment content is

- The most important part of any online presence
- Able to be easily aggregated, categorized and sorted
- Sharable via varying channels and formats
- Often generated directly by the user
- increasingly highly visual, with less text and more images and video. Users may see information aggregated in a personalized site, updated by a syndication

system, and tagged, reviewed or updated by other users. The web is now often mapped by 'tag clouds' in which masses of relevancy-weighted information surround an initial point of interest.

The organization and structure of the web is largely defined and managed by its users. ' Tagging' means applying labels to web content that describe it. This makes them searchable and sort-able by users. Tags can be applied by both suppliers and users- if a site allows them to. Popular Web 2.0 sites such as YouTube, Flickr and Gmail, work on user tagging systems. Increasingly, tagging and recognition of users identities via cookies means that XML websites are increasingly tailored to the users and that users each have a completely different experience of the website, making it their version of the site.

This environment of manageable, categorized content also allows XML to facilitate data exchange- which means that we can combine data from different sources into a single web page or application. This process is known as 'Mashing' and has become very important to tourism by allowing user generated content (UGC) to be embedded into supplier sites. The mashing of tourism content and links (e.g. cultural contents and links) into online maps has created huge opportunities for the promotion and management of tourism resources in a highly visual and interactive manner.

The Cloud GIS marketing becomes widespread, many destinations and suppliers are forming partnerships with UGC aggregators and publishers to share and mash content. This content plays a major role in awareness and selection of destinations and travel suppliers, and is a key element in the distribution chain. With this decision made, web users are now using the web to preserve and valorize cultural and natural heritage.

Achieving this means having XML enabled supplier sites or destination management systems that can provide realtime access to inventory of rooms, seats and travel product, supported by the ability to transact via e-commerce.

Heritage tourism, as a part of the broader category of "cultural tourism", is now a major pillar of the nascent tourism strategy of many countries. Cultural heritage tourism strategies in various countries have in common that they are a major growth area, that they can be used to boost local culture, and that they can aid the seasonal and geographic spread of tourism (Richards, 2003).

Cultural tourism can be defined as visits by persons from outside the host community motivated wholly or in part by interest in the historical, artistic, scientific or lifestyle/ heritage offerings of a community, region, group or institution (Silberberg, 2003) while heritage tourism is a broad field of specialty travel, based on nostalgia for the past and the desire to experience diverse cultural landscapes and forms. It includes travel to festivals and other cultural events, visit to sites and monuments, travel to study nature, folklore or art or pilgrimages (Zeppel and Hall, 2002). The word "heritage" in its broader meaning is generally associated with the word "inheritance." that is, something transferred from one generation to another. Owing to its role as a carrier of historical values from the past, heritage is viewed as part of the cultural tradition of a society. The concept of "tourism," on the other hand, is really a form of modern consciousness (Nuryanti, 2004). In this research, both heritage and cultural tourism are used in combination and or interchangeably with the explotation of web technologies. Recent studies about cultural heritage tourism and new information and communication technologies have focused on identifying the characteristics, development, and management of cultural heritage tourism, as well as on investigating demographic and travel behaviour characteristics of tourists who visit cultural heritage destinations. Micelli and Di Maria (2000) provided a common pattern of cultural heritage towns by analysing the way in which digital technology have significally changes consumer's demand. Formica and Uysal (2000) explored the existing markets of a unique annual event that blends internationally well-known cultural exhibitions with the new information and communication technologies. Behavioral, motivational, and demographic characteristics of web visitors were examined by using a posteriori market segmentation. Pilati (2002) explored the way in which Web technologies can create new opportunities and tools of exploitation of cultural heritage together with other technologies in order to provide information and knowledge to art consumers.

Musumeci (2000) also analysed how that the exploitation of cultural patrimony produces a positive impact on the economy, particularly in those areas that are better equipped from a cultural and tourist point of view.

ICOMOS analysed how the Preservation planning is perhaps the most effective use of some particular technologies such as geographic information system, since careful planning can prevent some threats to Cultural Heritage sites before they become problems.

According to the Charter on Cultural Tourism (adopted by ICOMOS): "Representatives of these bodies [ICOMOS]... express the wish that the states... shall adopt all appropriate measures to facilitate the information and training of persons travelling for tourist purposes inside and out of their country of origin."

In conclusion we can say that Information Technology (IT) provides modern means of distributing information to the mass in order to lead to a better-informed public debate. Information has to be prepared and presented in an intelligible manner for the citizens who do not have a technical knowledge of the process. This is especially true in modern societies in which IT is part of everyday life. According to Shiffer (2002), "information technology can help people to comprehend information (cultural information), thereby delivering knowledge". The main categories of information displayed and used in this section, and the descriptions accompanying them were also implemented by a team of Web design of Consortium Forma's experts based on common knowledge of the Cultural Heritage Institution (by a team of scientific consultants (both internal and external) of Consorzio Forma, each one with diverse approach and field of interest and studies – from archaelogy to modern or contemporary art, from high education to cultural heritage management, from web design and data modeling to quality control of web sites accessibility). Based on these considerations the project goal was to develop a single, operational and technical prototype (software and services) which can be set up as a cultural portal offering to visit on the internet sources on number of museums, archives, libraries, archaeological sites, digital cultural heritage, high quality agriculture, food specialities production (http://www.parks.it/parco.etna/pro.html), etc. It can work as a joint cultural portal (horizontal portal) and as a search engine. The platform should be compatible with the "Scegli Italia" portal of the Italian Ministry of Culture (MiBAC) in order to implement the Innovation Information Technologies and the Tourism market (Lo Tauro, A. 2003, and Lo Tauro, A. 2012). The system should be able to easily produce several services such as real-time thematic maps, more or less complex queries and visualizations, in order to implement tourism on-line services and cultural promotional tasks.

2.3.1 Preservation/Valorisation of Heritage with GIS technologies for CT

One of the most important uses of a GIS for the Cultural Heritage would be in the planning of tourism. Cultural tourism is both a friend and for of historic and natural sites. The money it brings in can be used to support the management of the site and to benefit the surrounding community. In 2000, there was an estimated $3.1 trillion of revenue generated by tourism, and 130 million people were employed in this industry. The World Tourism Organization (WTO) forecasts that by the year 2004 there will be 661 million international (non-domestic) tourist arrivals, representing an annual growth of 3.8%.

ICOMOS and UNESCO are supportive of tourism to Cultural Heritage sites; in fact, the very idea of the World Heritage implies public access to these sites. On the other hand, both organizations are concerned about the destructive effect which tourism can have on a site. Tourism can result in increased air pollution, noise, lights, new development near the site, wearing of stone floors and steps, erosion, damage from vehicles, vandalism and theft. According to an article in the *UNESCO Courier* from 1996, UNESCO encourages study and action in four areas:

- defense and promotion of the cultural heritage and cultural identity;
- protection of the natural environment against inroads from tourism;
- encouragement of a form of tourism that respects the social, cultural and natural environment and provides a basis for development;
- extension of cultural tourism activities that generate income and jobs, especially for young people in developing countries. (Bequette, 1996)

UNESCO and ICOMOS are not alone in their concern for sustainable tourism. In 1995, 500 participants in the World Conference on Sustainable Tourism adopted the Charter for Sustainable Tourism. This Charter states, "Tourism should contribute to sustainable development and be integrated with the natural, cultural and human environment; it must respect the fragile balances that characterize many tourist destinations, in particular small islands and environmentally sensitive areas. Tourism should ensure an acceptable evolution as regards its influence on natural resources, biodiversity and the capacity for assimilation of any impacts and residues produced."

States Parties are encouraged to document and protect all culturally and naturally important sites within their borders, not just those which are already listed as World Heritage sites. On such a large scale, tourism planning can play an important role in the protection of sites. For example, some sites are over-visited while others have not yet been discovered by tourists. One way to protect over-visited sites is to divert tourism to other sites. Which sites to develop for tourism should be decided not only on the basis of the sites' cultural or natural interest but also on their ability to support tourism.

For both natural and cultural sites, a tourism carrying capacity can be calculated using a GIS. This analysis would be based on such factors as cultural itineraries can valorise the environment in related to soil type, biological fragility (some species can tolerate more human interaction than others), cultural sources, fauna, etc. Available water and parking space would be considered, with a goal of minimal intervention. Other factors to consider would be the proximity of existing infrastructure such as roads and airports, and the proximity of other sites of interest. Spatial elements such as these are important to consider not only in the selection of a site to develop, but in the planning of that site. A large natural area may have many possible entry points, but one may wish to encourage entry in a particular location because of its lower fragility and proximity to other sites of interest.

A model for the selection of a tourist destination could include the elements shown in table 2.1.

These lists could be expanded upon, but considering only the elements listed above would result in a complex model. To perform all of these analyses with paper maps would be extremely time-consuming. With novel databases, the task of analysis is much simpler (Vanschoren, J., et. Al. 2012).

Analysis commonly begins with an operation to remove unsuitable sites from the analysis, those sites which do not meet the boundary conditions. Then the remaining sites are rated in a multi-step process and spatial aspects are weighted according to their importance. For example, one might first remove from the study area any land that falls outside the country's borders. Then one might query the GIS to show the 100-kilometer radius with the highest density of cultural and natural sites, or to show the top five areas in order of density of potential tourist attractions. Having located clusters of possible tourist destinations, a GIS can facilitate the rating of sites according to their cultural, ecological and economic suitability and their projected number of tourist arrivals.

Some types of analysis, such as classifying monuments by age and type, can be done statistically with the results displayed as text and numbers. A GIS does more: it allows the user to see spatial relationships such as clusters of sites, patterns of distribution (linear groupings which may suggest a tourist route rather than a single destination), and relationships to other elements which may not have been clear in the initial formulation of the model. Spatial models usually include the option of refining the model at several

Cultural Elements	**Natural Elements**
Favorable: Proximity to roads	Soil Type
Type of road (capacity & speed)	Steepness
Proximity to airports	Aspect (e.g. north-facing)
Capacity of airport	Biological Fragility
Proximity to border Proximity to population centers Size of population center	Proximity to wild populations (e.g. large mammals or bird colonies)
Proximity to cultural monuments	Proximity to special geographic points (e.g. beaches, caves)
Importance of monument	Importance of geographic point
Unfavorable: Visual access to distracting views	Visual quality (views)

TABLE 2.1: CULTURAL ELEMENTS FOR TOURIST DESTINATION

stages, because the analysis process usually reveals erroneous assumptions. It is the automatic and rapid display of selected spatial data as a map which allows a researcher to explore many alternatives and to perform a complex analysis in a reasonable time period.

The Charter for Sustainable Tourism states, "all those responsible" must take upon themselves a true challenge, that of cultural, technological and professional innovation, and must also undertake a major effort to create and implement integrated planning and management instruments." A GIS is one sort of innovation which can help those responsible for planning sustainable tourism.

2.4 Remote Sensing (RS)

Since their inception, RS technologies have provided researchers and practitioners with the ability to produce land cover mapping. The first tangible combination of space exploration and remote sensing took place after World War II when cameras were attached to rockets and launched from New Mexico (Lillesand and Kiefer, 1994). These images were not of high quality, but they did initiate the interest of researchers in taking pictures from space. Following this, in the early 1960s, weather satellites captured images of clouds with the coarse earth's surface as the background. Since then, many satellites with various sensors have been put into use by many countries for the general purpose of earth observation.

Two of the most successful and influential satellite programs include the Landsat program in the United States and the SPOT (Systeme pour l'Observation de la Terre) program in France. The first Landsat Mission was in 1975 and since then seven missions have been completed, all but one successfully. The SPOT program has produced four satellites since 1986 with a fifth scheduled for launch before the end of 2002. Both the 24 Landsat and SPOT programs have improved and refined their data collection methods since the original launches and today are amongst the leaders in breadth of wavelength coverage and improved image spatial resolution. Most recently, the launch of the private satellite IKONOS by Space Imaging Inc. in 1999 marks the availability of the world's first commercial high-resolution (< 2 meters) satellite imagery (Tanaka and Sugimura, 2001). Also, the launch of QuickBird on October 18th 2001 by Digital Globe, which offers resolutions similar to IKONOS, will bring competition to the high-resolution commercial satellite market.

Morain (1998) lists the following items as the motivational forces behind the development of remote sensing technologies:

1) need for better information about the earth's surface

- spatial distribution of natural resources
- effective management of renewable and non renewable resources

2) security

- assets for maintaining security through observation

3) commercial opportunities

- still being tested via recent launches of privately owned satellites

4) international co-operation

- co-operation to better understand, manage, share, and protect earth's resources

5) international law

- to ensure the facilitation of free and open exchanges of data and information.

These points outline the main uses of data collected via RS since its inception and subsequent development. These uses are applicable to planning and policy making needs in parks. Because RS can provide land cover data relatively swiftly and efficiently, it can be concluded that it is a suitable vehicle for planning and the management (Ulbricht *et al.*, 2001) of environmental sources.

Using RS, planners can better understand the environments they work in, thereby assisting in problem solving and decision-making processes. In particular the GMES (http://www.gmes.info/) requires in the long-term the deployment of a comprehensive and complementary set of operational space missions providing, at global / regional / local scales, permanent and continuous observing capabilities of the Earth's system components namely, the the land surface, also including the atmosphere, the oceans, and the ice masses. In order to have the operational space elements in place as from the beginning of the full operational phase (2007-2013), the preparatory work started in 2004.

As anticipated in the European Space Policy White Paper under GMES heading, there will be different series of satellites, each series covering a given domain (high-resolution imagery, wide field imagery for ocean and land, etc.).

These systems provided long-term continuous access to the following measurements:

- Very high (1 meter or better) to high (10 meters) spatial resolution, panchromatic and multi-spectral optical imagery for local observations over land surface and coastal zones.

- High (10 meters) to medium (200 meters) spatial resolution, wide field-of-view, multispectral and multi-directional optical imagery for global / regional observations over land and ocean surface. This is needed in support of global vegetation (including forestry) and biosphere monitoring (including coastal zone water quality management, coastal surveillance, disaster management and humanitarian aid);

- Multi-spectral optical imaging satellites at two spatial resolutions:

  - High resolution for local and regional operational monitoring applications (continuity of SPOT and Landsat classes)

  - Medium resolution for global applications (continuity of ENVISAT-global imaging and SPOT-Vegetation -classes), with multi-spectral capabilities and optimized for vegetation.

In this context, Da Costa and Cintra (2001), in their paper addressing the use of RS for environmental analysis, note the abilities of SPOT and Landsat imagery in the areas of research support, analysis of spatial dynamics and urban morphology. Similar studies were conducted by Nocentini, (2004) in parks and natural reserves in Italy. They further note that the methodology, which includes the use of RS analysis, permits the study of environments in a complete, rapid and precise manner, all at a relatively low cost (Da Costa and Cintra, 2001,).

RS has been extensively used in the study of land cover and land use change across the globe. While this work has had the common goal of facilitating identification, classification and changes in the distribution of land cover and land use, approaches and data sources have varied. Land cover can be defined as the physical composition of the land under examination, or what can be recognized from the air, while land use is the actual usage or productive application of the land. For example, land cover could be forest, grassed vegetation, or water body whereas a land use would be agriculture, infrastructure, cultural heritage etc. Because of the many satellites and sensors that are currently orbiting the earth, researchers have several options when it comes to selecting the sensor that best targets the location and needs they are interested in. For example, Martinez-Casasnovas (2000) implemented a cartographic and database approach for land cover/use mapping in Spain. His approach began with the development of a hierarchy of terrain object classes to support classification results. The results of this hierarchy were then transformed into basic mapping generalizations, which were in turn translated into a relational database focused on structuring data in order of data element relationships. He concluded that this approach to image classification can be viewed as advantageous from the user's point of view since land cover / use classes were structurally broken down into basic mapping units that could be used in a variety of planning applications.

Similarly, Makarovic (2003) discussed image-based mapping, listing image interpretation and feature extraction, stereomodel orientations, sampling and digitizing, geometric transformations and field completion as the five main mapping techniques. The technique of particular interest in his paper was image interpretation and feature extraction. Makarovic indicated that feature extraction was easier and more effective when high-quality (higher resolution) images were utilized. Because of this, image-based mapping was best suited to semi-detailed and coarse thematic mapping, where high accuracy was not required. This continues to be the case with even higher resolution satellite imagery. Not all RS image analysis involves straightforward classification processes. Several computing algorithms including edge detectors, neural networks, and dynamic programming techniques have been developed to analyse land cover with RS data. For example, Kalluri et al. (2000) used a hierarchical connected components algorithm to undertake image segmentation by clustering pixels into homogeneous regions. They then connected these components relative to uniform pixel reflectances. Kalluri et al. (2000), conclude in their discussion that recent developments in faster, high-end servers and workstations allowed the development of complex mapping and computational processes that were cumbersome to implement in the past. They recommend the use of multiple processors in data processing for the computationally intensive tasks of image enhancement and segmentation and they noted that new processing technologies will ensure the implementation of remote sensing algorithms to study land cover dynamics at a global scale (Kalluri et al., 2000). The research is oriented to the design of novel algorithms (Barranquero, J., González, P., Díez, J., José del Coz. J., 2013)

Entwisle and Walsh (2004) suggest that RS data can provide a valuable perspective on land cover and land use change. Using multi-source and multi-temporal RS imagery, Entwisle and Walsh (2004) were able to examine inter-seasonal variations of the landscape that were otherwise difficult to assemble and acquire. The data allowed enhancement of the overall perspective of their study when integrated with the ground-based population surveys. However, for temporal analysis, several limitations affect remote international locations. For example, archived data are often difficult to find, assemble and acquire and few alternate views of the landscape are offered from the commonly available satellite systems (Entwisle and Walsh, 2004).

Since the early 1990s, the technique of SAR Interferometry (InSAR) has been shown to lead accurate large-scale surface displacement mapping for various phenomena (Massonnet and Feigl, 1998).

In a related study, Dutra and Huber (1999) examined feature extraction for land use classification using Earth Resources Satellite Synthetic Aperture Radar (ERS SAR) data. They confirmed the notion that specific feature extraction in a land cover classification is inherently difficult given that it is both dependent on the input data and on the classifier used. Their method involved feature selection followed by classification using the maximum-likelihood classifier and an artificial neural network. Because of their use of SAR data, they were forced to employ a refining Lee filter to remove the speckle inherent in the image, a technique often employed with SAR data (Hall *et al*, 2001). Ultimately, Dutra and Huber (1999) found that their classification accuracy, similar to most classification methods, was excellent in terms of the training set, but worse for the validation set and subsequently for the complete image.

Advanced spatial technologies such as photogrammetry and lidar have improved the quality of spatial information and enable data processing for more accurate estimation of environment parameters. This research aims to develop a quantification method for sustainability indexes by using spatial metrics such as compactness, complexity and density. Although building height information is an important element of urban morphology, it has been neglected in previous studies. Hence, height information obtained by lidar is incorporated into the spatial metrics in this study. The spatial metrics are applied to four study cases. We have examined the metrics and concluded that the developed metrics can quantify the sustainable urban form concept more effectively. The main finding of this study confirms that the 3-dimensional spatial metrics differentiate the complexity of areas significantly. Another significance of this study is the high capability of spatial metrics for the quantification of sustainable urban forms in terms of complexity, compactness and density. The developed indexes can be used for the determination of the spatio-temporal changes of sustainable forms or the comparison of the cities in terms of a sustainable urban form using remotely sensed data (Shirowzhan, S. and Lim, S. 2012).

2.5. Geographical Information Systems (GIS)

In the early 1990's, spatial information was used by public land managers to estimate the forest types and forest distribution in the US (Graetz 1990, Lannom 1995). It was also used to produce thematic mapped data (Evans 1994) and mapped forest density (Zhu 1994). GIS has proved to be a success in Environmental Impact Assessment (EIA), a procedure becoming mandatory for any large project in many places in the world (Johnston et al. 1994). It has been used in a wide variety of natural resource management practices (Schreier and Brown 1992). It has been used in evaluating damage from natural disasters (Anonymous 1992, Tsuji et al. 1996), and in monitoring pollution contamination of various types (Runyon et al. 1994). GIS have been integrated with knowledge bases to produce expert systems for natural resource management (Ji et al. 1992). It was widely used for National Parks and wildland protection, management, and planning (Wilcox 1984, Spear and Cottrill 1993, Murphy and Smith 1995, Niemann and Niemann 1996). It has been used widely in environmental monitoring (Ramsey et al. 1994, Johnston and Ji 1994), e.g., monitoring land degradation and soil loss (Roehl 1962, Wischmeier and Smith 1968,1978; Wade and Wickham 1995).

Additional GIS use has been as a tool for data collection about habitats and forming a habitat data bank (McGarile 1994, Thompson et al. 2000). GIS were used in landscape architecture, urban and regional planning, and studying regional growth (Ehlers et al. 1990, Bou-Saab and Steingraber 1992, Treitz et al. 1992, Genrich and Cooper 1993, Azar et al. 1994, Bocco and Sanchez 1995). GIS were used also in finding different alternatives for projects planning and in visualizing the future state of a study area (Versenyi et al. 1994, Batty and Howes 1996). The military have intensively used GIS for 30 years (Kish 1995, Shank 1996). Lately the military has used GIS to find other uses for military lands (Sherman 1995). GIS has been used in cultural heritage studies around the world, revealing hidden cultural and historical sites (Lopata et al. 1992, Murphy and Smith 1995). GIS has now entered many businesses, connecting information bases with branches on regional, national, and international scales (Corbley 1995, Longdom and Barnes 1995, Chalmers 1996, Chisholm and Brown 1996). The planning of some businesses has now became dependent on GIS applications (Benoit 2001, Thrall 2002, Thompson et al. 2000). Park GIS are able to relay both management needs of related protected areas and environmental control for the integration of environmental assessment (Spina S. 2004)

The framework encompassed multiple criteria and included assessment using GIS, and in particular, web-based Geographical Information System (GIS) tools. Study sites were chosen due to its unique environmental characteristics and its proximity to allowing easy site visits and data access (Ruza, J., Kim, J., Leung, I., Coyan, C., Li, A., Kam, C., and Ng, S., 2012)

2.5.1 A Brief Description of a GIS

Aronoff (1989) defined the GIS as any computer-based procedures to capture and manipulate geographically referenced data (spatial data). Dueker and Kjerne (1989) defined GIS as " a system of hardware, software, data, people, organizations, and institutional arrangements for collecting, storing, analysing, and disseminating information about areas of the earth." The basic function within a GIS is to divide a map into a group of small single factor units. McHarg (1969) in *Design with Nature* was one of the first to popularise overlay mapping in the context of land use planning. His book appeared when the early computer mapping became readily available. Overlay mapping with physical map products has been widely used; it can now be used in many ways and has new expanded practical scope due to computer and printer advances. The entire matrix of values about one topic or factor on a map is a "layer." By putting the layers together, the original map can be re-created. Each single-factor map may be revised to be more (or less) precise depending on the project or use. In raster-based works, the smaller the cell, the more precise will be the details. Very frequently, 30 x 30 m is used for a cell size. I used 30 x 30m cells in my demonstration. This scale has the same size as the Landsat Thematic Mapper sensor resolution, a resolution used for the land-use classification described later.

For every land feature there must be within a GIS the name, exactly where it is, and how it relates to other features. This information about the recorded features include position, topographical information and spatial relationship, and feature attributes.

Acquiring and maintaining the data sets are the most expensive component of any GIS.

Between 60% and 90% of the investment in a GIS relates to data sets. The data about land features are entered by scan digitizing, manual digitizing, manual keyboard data entry, and by using procedures available with softwares used with the global positioning satellite system (hereinafter GPS). The spatial data recorded for a feature could be in the form of vector or raster layers. Vector data are representations of polygons. Raster data, in the form of a matrix of values, describe conditions or features at points, typically the centre of the cells of a matrix. The GPS may be used to determine and record the exact position of a feature or observation relative to a global division. About 50 years ago, world map makers devised the Universal Transverse Mercator grid (hereinafter UTM). Using the UTM coordinates with the GPS seemed best suited to overcome the confusion in using many coordinate systems. Using maps without relating to the National Cartographic Grid System becomes like using a street map without streets names (Neri 1996).

Different methods have been analysed to segment laser scanner data in order to extract information and in order to obtain topographical and object height information for mapping and GIS tasks. For this study, laser scanner data and scanned topographical maps are to be used. Firstly, the region-based segmentation method is used to delimit objects. Then a scanned topographic map, filtered and converted into a vector format, is utilised to locate already recorded greenways and cultural heritage. Greenways not contained in the map could be found by combination of segment attributes.

GIS is an accumulative system which can be used to satisfy human needs to know what exists at a particular site, what conditions have changed over time, where it changed, and what might happen if the existing use is altered. Any quantifiable social, economic, or environmental impacts due to altering some of the conditions or uses, potentially may be monitored for a specific site using the GIS (Walker and Miller 1990).

A GIS is often used to produce a visual image of an area of study. The graphic side of a colorful spectrum of raster and vector layers of GIS may have more impact on viewers than other kinds of automation or modeling (Parr 1992). An advantage of using a GIS lies in the ability to use visual layers that then may be overlain or combined in several ways with other aspects of the study area (Giles 1962,1985). Images from satellites, infrared photos, and radar were widely used in remote sensing mapping (Collwell 1973). GIS application followed such use in producing images of layers, many of which were intended to be helpful in planning (Fuller 1994).

In the beginning of the 1900's, many problem faced those who tried to draw an accurate, reliable map. The most complicated problem was to collect the needed information from different maps having different scales. A group of bays and fjords of a coast line may be represented by a straight line in a less detailed or bigger scale map. That lead scientists and mathematicians worked hard on finding a process by which detailed relations may stay undistorted throughout scale changes. Some mathematicians called this process scale-linking procedure. That procedure consist of direct consequence of identical processes shaping land form across many scales. The process is also known as "fractal scaling" which is widely used by geologists, landscape architects, and ecologists.

(Van der Ryn and Cowan 2002) claimed that unless people start working with nature's own finely-tuned, scale-linking systems, they will be endangering the stability of the planet. Although this statement may be frightening, I agree that a link will have to be established to use previous efforts that were under taken in ecology at different scales.

Another concept to be considered is one of not focusing on a single scale but finding an interactive link between different scales. Failing to find this link leads managers to miss the opportunities to work across different scales in a unified method to address a problem. GIS could be our way to this kind of interactive link, a pathway for studying an area as a group of adjacent cells. By unifying the size, shape, and scale of the cells, GIS users can then study an area on different scales depending on the number of cells to be considered as a study area. By judicious use of available software, it is possible to relate information in a map or matrix of cells to factors in adjacent or nearby cells. Juxtaposition is a very important factor in ecological studies. It mainly depends on the quality of relations between adjacent areas. To study the ecological system of any area we have to know the answer to some important questions, such as, how far the site is from water; how close an area or point is to edge; how near a site is to a greenway; or how far a site is from another site of a specific property. Novel implementations can answer these questions on visual maps (Abe: A. 2012, Lo Tauro, A. 2012).

2.5.2 A Brief Description of a Cloud GIS

GIS Cloud is the first purely web based GIS powered by cloud computing. It provides full desktop GIS features enriched by the web. GIS Cloud offers easy and efficient visualization, analysis and exploration of geographic information. The primary goals of the GIS Cloud platform are as follows; to simplify exchange of geographical information between users and offer an easy way to analyze this information regardless of the location of its users. By using GIS Cloud our users access the full power of desktop GIS, allowing for such activities as geospatial analysis, spatial intelligence, the creation of customized mapping reports and publishing geographic analysis on the Web.

The use of open source software, cloud computing technologies, and an integrated approach to search, video, and media seems almost common-place in industry these days. Yet government websites aren't quite there with the exception of a few noteable exceptions (not an exhaustive list by any means).

With GISCloud it is possible to create a very wide range of GIS projects and analyses. Shown below is just a research possibility. We developed a GISCloud for the blue infrastructure to improve management of canals. This has also had a large effect on the efficiency of emergency interventions and maintenance.

Gain huge cost savings by learning how to leverage the power of both laser scanning solutions, including CloudWorx softwares, and Smart 3D technology (SmartPlant 3D and SmartMarine 3D). We tested solutions to manage costs and ensure accuracy on the project (http://blogs.intergraph.com/getsmart/).

The upcoming version of CloudWorx for Smart 3D, helped me to:

* Navigate point clouds in the Smart 3D graphic environment easier and faster with additional view controls

* Use point cloud data to interact with and create/modify native Smart 3D models

* Read and manipulate point cloud data from any scanning source

FIGURE 2-1-2: GIS CLOUD FOR THE BLUE INFRASTRUCTURE

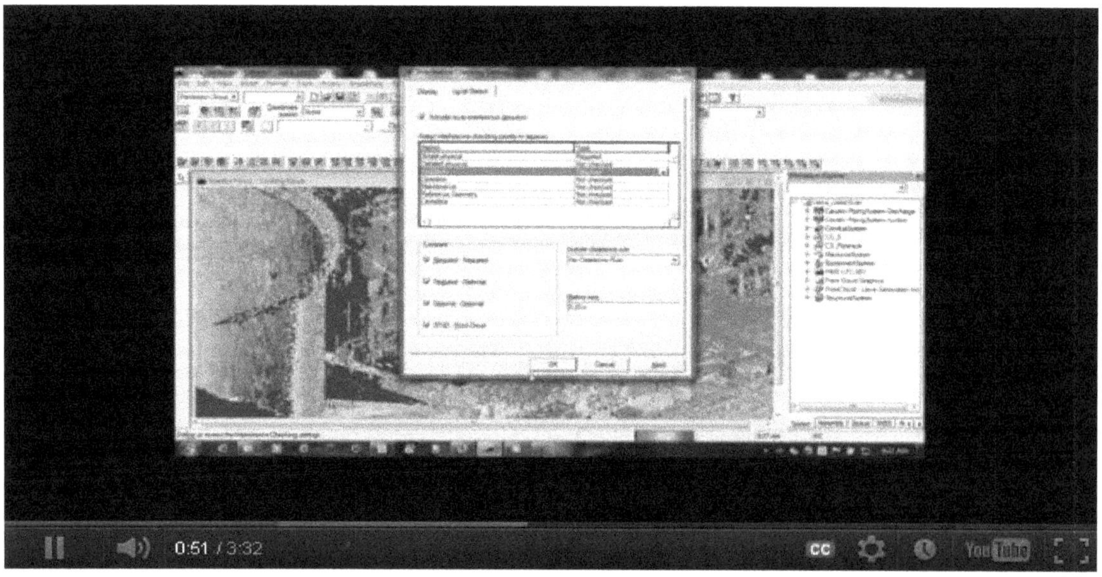

FIGURE 2-1-3: LASER SCANNING VIDEO WITH SMART 3D

2.5.3 Intelligent location and data

The role of location intelligence plays an important role in critical decisions. Softwares as a Service is the latest GIS innovation to make "the jump from trial to trusted". Cloud computing describes an approach to delivering solutions where provides host and manage software and associated data on behalf of their clients. An explosion of GIS applications along with expanded user populations have added pressure to already-strained IT resources. The sheer amount and growing complexity of location data and images necessitate increased capacity and bandwidth. Combined with heightened expectation for speed, many organizations have needed to employ aotdated data, inefficient workflows or costly solutions. Cloud computing has changed the dynamics.

Location-based information involves:

- Lower total cost of ownership
- Qicker time to value
- Greater resource utilization
- Continuous improvement
- Intelligent collaboration

IT managers play an important role in assessing the realibility, stability and security of any off-site hosting platform. Cloud computing is an approach to information processing where computing technology, including softwares and data (including open data) is provided as service. In 2009, this method of delivery reached critical mass and accounted for more than 15% of all new software installations.

The NATO Open Source Intelligence research is one of the standard references available to the public. The other two NATO references are the NATO Open Source Intelligence Reader and the NATO Intelligence Exploitation of the Internet.

On demand applications now add value across a broad range of business, but are particularly relevant in environments that relay on spatial data and location intelligence where managers must deal with issues around growing complexity, data security and cost management.

The one thing certain about the future of data is more data and the need for location intelligence grow exponentially. With the proliferation of oriented mapping applications and more intuitive interface designs, company need to put geo-analytical tools in the hands of an increasing number of users. They need more open data. *Open data is data that can be freely used, reused and redistributed by anyone - subject only, at most, to the requirement to attribute and sharealike.*

The full Open Definition gives precise details as to what this means. To summarize the most important:

- Availability and Access: the data must be available as a whole and at no more than a reasonable reproduction cost, preferably by downloading over the internet. The data must also be available in a convenient and modifiable form.

- Reuse and Redistribution: the data must be provided under terms that permit reuse and redistribution including the intermixing with other datasets.
- Universal Participation: everyone must be able to use, reuse and redistribute - there should be no discrimination against fields of endeavour or against persons or groups. For example, 'non-commercial' restrictions that would prevent 'commercial' use, or restrictions of use for certain purposes (e.g. only in education), are not allowed.

If you're wondering why it is so important to be clear about what open means and why this definition is used, there's a simple answer: interoperability.

Interoperability denotes the ability of diverse systems and organizations to work together (inter-operate). In this case, it is the ability to interoperate - or intermix - different datasets.

Interoperability is important because it allows for different components to work together. This ability to componentize and to 'plug together' components is essential to building large, complex systems. Without interoperability this becomes near impossible — as evidenced in the most famous myth of the Tower of Babel where the (in)ability to communicate (to interoperate) resulted in the complete breakdown of the tower-building effort.

We face a similar situation with regard to data. The core of a "commons" of data (or code) is that one piece of "open" material contained therein can be freely intermixed with other "open" material. This interoperability is absolutely key to realizing the main practical benefits of "openness": the dramatically enhanced ability to combine different datasets together and thereby to develop more and better products and services (these benefits are discussed in more detail in the section on 'why' open data).

Providing a clear definition of openness ensures that when you get two open datasets from two different sources, you will be able to combine them together, and it ensures that we avoid our own 'tower of babel': lots of datasets but little or no ability to combine them together into the larger systems where the real value lies (http://opendatahandbook.org/)

2.6 Integration of Remote Sensing and GIS

SIT technologies, especially RS and GIS, have much to benefit from each other (Foody, 2005 and Rozenstein, O. Karnieli, A. 2011).

In fact, some software vendors, such as ESRI and ERDAS, claim the ability to provide seamless integration between products (http://www.planetek.it/). However, practice has shown that the coupling of RS and GIS data formats is still some way from seamless. The integration is hampered by the use of differing data models and by the historical differences between these two data structures (Edwards, 2001). Data models in this context include raster, which represents features in a cell or regular grid based model, and vector, which uses points, lines and polygons to represent features. Total SIT integration is defined by three classes of integration, namely 'separate but equal', 'seamless', and 'total integration' (Edwards, 2001). Separate but equal integration allows for separate software interfaces for image analysis and GIS with data exchange between each of the two, while seamless integration combines raster and vector systems into a single interface. Total integration involves a single system without the encumbrances of classes one and two noted above (Edwards, 2001). Several researchers have recently analyzed the current state of successful integration, at various levels, between RS and GIS (Gahegan and Ehlers 2000, Edwards 2001). Wilkinson (2004) proposed three main ways in which RS and GIS technologies are complementary to each other:

1) remote sensing can be used as a tool to gather datasets for use in GIS,

2) GIS datasets can be used as ancillary information to improve products derived from remote sensing, and

3) remote sensing data and GIS data can be used together in environmental modelling analysis.

The use of RS as a tool to gather datasets for analysis in a GIS is not new. However, improvements in image classification are making their integration with ground-based data more successful than ever before (Hall *et al.* 2001). New classification algorithms and innovations such as artificial neural networks are improving classification methods and results and trends in gathering and integrating analysis of RS and GIS data suggest a convergence of the technologies. Wilkinson (2004) suggested attribute analysis and classification of RS and GIS data, the use of very high dimensionality data from RS and GIS, and remote environmental mapping and object analysis/search in integrated RS and GIS datasets were commonplace analyses reported in the literature. However, within the field of remote sensing for mapping purposes, the fusion of multi-sensor, temporal data, while practiced for years, is still a complicated process.

Despite the advances, the processes involved in data fusion are still currently under development as much as at any point in the past decade. This situation is somewhat complicated by the constant introduction of new and different satellite remote sensors. Moreover, many types of different formats and structures continue to hamper this integration. GMES (www. http://www.gmes.info/ and http://europa.eu.int/comm/space/programmes/gmes_en.html) is the major European contribution to this goal, but at the same time must satisfy the specific needs of European policy-makers. Needs in support of external aid and european policies under the GMES programme are identical to some needs for environmental monitoring such as improved access to Earth observation data, sustainability of the data, and better integration of Earth observation data with background data.

The need for international co-operation was recently reinforced at the highest political levels at the World Summit on Sustainable Development (WSSD). At these events, the heads of state called for improved global observations for better decisionmaking, underscoring the critical link between global observations from space, airborne and in situ platforms.

The G8 Summit in Evian, June 2003, resulted in the G8 Action Plan on Science and Technology for Sustainable Development. The Summit expressed clearly the need to develop close international co-ordination of global observation strategies for the next ten years. This includes the need to identify new observations to minimise data gaps by building on existing work to produce reliable data products on atmosphere, land, fresh water, oceans and ecosystems.

There is also a strong need to improve the gathering and integration of many different types of information. For example, standardised mapping - an essential element for planning crisis management – is not readily available in up-to-date digital form for many of the EU's areas of interest. The updating of traditional hardcopy maps is very costly and requires highly trained cartographers. Photo-maps, on the other hand, can provide a cost-effective and relatively fast alternative, and commercial satellite imagery is well suited to the production of such items. Geographic information systems combined with a wide range of attribute data, such as data on population density, age distribution, health, infrastructure, land use, etc. are extremely useful to planners and decision-makers who need to have a global picture of the situation as it.

The updating of GIS data layers based on digital remote sensing has been extensively examined (Makarovic, 2003; Martinez-Casasnovas, 2000). Stow *et al.* (2001) introduced a process that a regional planning agency can use to correct and update vector data using remotely sensed data. Using what they dubbed a "live link" between ERDAS image processing software and ArcGis software, remotely sensed data were successfully adopted to correct and update GIS land use layers. In the years since their research, software vendors have gone to great lengths to facilitate integration without the need for a "live link" between two separate interfaces. For example, many GIS currently support both raster and vector data structures and several image analysis packages promise a single interface capable of various merged functionalities, architectures, data structures, and ultimately, user experience.

In the context of this project, the integration of RS and GIS technologies is facilitated as follows. RS imagery is proposed as a means to collect temporal land use change information. The results of this change detection form a dataset that can then be fed into a GIS for further integrated analysis. The temporal GIS-based data analyses are proposed as a source of ancillary information to improve the output of the RS change detection results.

Together, the RS and GIS analyses are used to derive the location of areas of potential environmental risks for further analysis and subsequent policy and planning. This type of environmental modelling demonstrates precisely Wilkinson's (2004) and Rozenstein's (2011) several ways that RS and GIS technologies are complementary. Even with all of the improvements in the integration of RS and GIS data sets, some uncertainty still exists. Gahegan and Ehlers (2000), for example, suggested that transformation processes, which are integral in RS-GIS integration, contribute to the overall uncertainty that develops within the data. They expanded on this by noting that because data are abstracted from their 'raw' form to the higher representations used by GIS, they must be passed through a number of different data models using a series of transformations.

Current approaches to RS and GIS integration typically involve several stages of transformation including image classification, object formation, and data structure conversion. Image classification represents the replacement of the 'raw' image data with thematic interpretations to assist in the understanding of trend structures or content of the imagery. This process typically occurs on a per-pixel basis, however some pixels may be left unassigned or grouped into an 'unknown' class. Object formation is the conversion of a thematic model into an independent entity with a defined shape. Errors can occur in the object formation transformation process due to positional uncertainty and error. Finally, in terms of data structure, GIS data formats are typically software-specific (Huxhold et al, 2000 and Rozenstein, O. Karnieli, A. 2011). Hence, to use data created or modified with one proprietary GIS always requires a data structure transformation for analysis with a different GIS. All of these transformations, alone or combined, can lead to uncertainty in the data. It is possible that data characteristics may change during the transformation and it is difficult to determine to what extent the transformation method used may have affected the data (Gahegan and Ehlers, 2000). These sources of uncertainty affect the completeness and consistency of data making them difficult to manage and account for during decision making. Given this, the current methodologies for RS and GIS integration are by no means perfect and the validity of results generated from processes must be carefully scrutinized.

Several studies have been undertaken to examine the integration of RS and GIS (Molenaar and Janssen, 2003; Amissah-Arthur *et al.*, 2000; Yeh and Li, 2001, Rozenstein, O. Karnieli, A. 2011). Subject to the caveats noted above, this integration has proven to be an effective and potentially powerful means of land use analysis and classification that can assist in the development of planning guidelines and policies.

Aerial remote sensing with MIVIS hyperspectral scanner within "L.A.R.A. Project" (Aerial Laboratory for Environmental Research) of CNR (National Research Council) started in 1991 following an agreement between the CNR the Ministry for the Development of Southern Italy in the context of the Strategic Project "Climate, land, and environment of Southern Italy" which aimed at developing a system for the acquisition from aircraft of multispectral imagery for remote sensing of environmental, land resources. MIVIS was also employed successfully to detect and to monitor Italian volcanoes (such as Etna) by the INGV (Rome) in collaboration with the CGP and Pollino National Park in Basilicata.

In particular over an area representative of the Mediterranean ecosystem was performed a preliminary study which aims at verifying the viability of the information derived from satellite hyperspectral data as to airborne data, recorded with different spatial and spectral resolutions. The area is located, in the Basilicata region, within the Pollino National Park, where satellite Hyperion and airborne MIVIS hyperspectral data were collected respectively on December 4, 2001 and November 9, 1998. The two data sets have been processed, according to the sensor typology, by calibration procedures such as radiometric, geometric, atmospheric and topographic corrections. They were then analysed by means of different supervised classification methods, after choosing training sets describing vegetation typologies and ophiolithic outcrops prevailing in the Pollino Park. The thematic maps so obtained allowed to depict the ecosystem variability with a comparable precision thus confirming the potentialities of satellite hyperspectral data for the precise natural monitoring (Cavalli, R. M., Fusilli, L., Pascucci, S., Pignatti, S. Poscolieri, M, 2005). This brief review suggests in general that the use of RS technologies, and in particular the MIVIS technologies in parks is viable, thus affirming the general approach adopted in this research.

Rozenstein, & O. Karnieli, A. (2011) noted that the full potential of using RS as a source for land cover data has been realized to date, particularly in terms of the integration of this data source with GIS.

The integration of RS technologies with ground-based GIS to analyze land use at the environmental level, specifically in order to detect land use change and subsequent indicators of potential environmental stresses, is now considered in more detail. The primary landscape feature used to identify new environmental sources in this research is the greenway network. Hence, research on the detection and extraction of greenways from RS imagery is reviewed in the following sections.

2.7 Indicators Derived from RS and GIS

From space, it is possible to see and study the earth as a complete organism, whose health depends on the health of all its elements (WCED, 1987). The main indicators of potential environmental stress as included in the operational portion of the figure 2-4, include land use/land cover and cultural landscape identification.

These three indicators are selected for use in determining the location of cultural heritage and cultural sources over time and, subsequently, their potential use in detecting environmental values. Based on these indictors, it is possible to classify land use in terms of the stages of development shown in Figure 2-4. The areas that pass rapidly from stage one to two or three can then be targeted for improved planning practice. Also, based on the integrated framework outlined in this chapter, these indicators play a key role in the RS change detection process used to localize areas of potential environmental stress.

If a rapid shift such as this goes unidentified, the probability of environmentally sound development occurring is unlikely. The conversion of agricultural land into urban land uses has become, and will continue to be, a serious issue for sustainable growth in parks (Li and Yeh, 2000 and Rina 2011).

This is particularly crucial in the Etna park.

The case study of change detection in cultural landscapes in the Etna Park were conducted to develop appropriate techniques for quantifying and analysing the landscape and territorial change since 1995. Methods were derived from cultural landscape change analysis, based on cadastral maps and land registers, and were rendered at a land plot level by the use of GIS. The database for the diachronic GIS, especially cadastral maps and land plot records, are discussed in detail in Lo Tauro e alt. (2005a, 2005b and 2012). This study emphasizes that an approach based on a land plot level and using a land record-based GIS provides valuable results relevant for planning processes and nature conservation in changing cultural landscapes. Specifically, inefficient urban sprawl makes ineffective use of land, by leaving large communication gaps between areas of development, and forces large-scale intrusion onto surrounding agricultural lands. In particular, fragmented

FIGURE 2-2: THE FIGURE INCLUDES LAND USE/LAND COVER AND CULTURAL LANDSCAPE IDENTIFICATION.

conversion of agricultural land to various disastrous human interventions, such as flow variations, excavation works and dike building can be harmful to biological conservation by inhibiting diversity of natural biological and wildlife habitats (Yeh and Li, 2001). Today, both afforestation and the natural process of secondary succession of fallow fields, meadows and pastures have created new and unusual sights in many places, sometimes negatively affecting the attraction of certain landscapes for tourists (Hunziker and Kienast, 1999) and in particular for cultural tourists. The analysis of cultural landscape change is generally based on a variety of sources, including topographic and historical maps, aerial and satellite sources, land registers with geodetic survey maps and land plot records, original surveys of relict species (where available), as well as various statistical and archival data. A state of the landscape, which represents the traditional land use system, should be the starting point for the diachronic analysis of landscapes. I assume that analyses of cultural landscapes ultimately require operationalisation with the help of a remote sensing technologies and GIS in order to successfully manage the abundant information about land units, attribute data, and temporal layers (Fig. 5). The GIS serves here predominantly to analyse the changes and to calculate the proportion of each land use type in order to analyse the temporal change detection (Lo Tauro et alt, 2005b and 2012). The Etna Park is characterised by four zones (D.P.R.S. n. 37, March 17, 1987,):

Zone A covers 19,237 hectares and is mainly made up of public land which does not support any human communities. It is a widerness area, kingdom of large birds of prey including the golden aùeagle, where only minimal human interaction is permitted.

Zone B covers 25,391 hectares and is mainly made up of small, privately owned agricultural plots. Examples of traditional farmhouses, bare shelters for animals, wine, oil presses and austere patrician villas are scattered throughout this area and show the presence of man (Failla et al 2005).

In order to prevent marring the environment and nature, additional pre-park area covering about 14,000 hectares has been set aside for specially designed, unobtrusive tourist amenities. The areas belongs to zone C (4.188 Ha) and zone D (9.551 Ha). At the D zone of the Etna Park of Figure 2-4, much of the land use is likely to be range or idle land, as intense agricultural use has been abandoned due to various disastrous human interventions, such as flow variations, excavation works and dike building. Indicators such as large tracts of continuous homogenous land cover and relatively large acreages still under agricultural land use are used to determine whether or not land belongs to the D zone. As the land use at the fringe changes from agricultural and idle uses into urban land uses, such as road development, greenways network and initial building stages, a shift from zone D to zone C is generated (Figure 2-4). In general, the more rapid the shift is from the D zone to the C zone, the more likely the potential for environmental stress to occur as there has not likely been sufficient time for proper environmental management polices to have been implemented by planners.

In particular the aim is to encourage the valorization of traditional activities in order to prevent these areas being swallowed up by cement and concrete.

The built-up areas of the communities are not included in the park, even if part of their territory lies within the protected areas. Territory of the following communities is included in the Park: Adrano, Belpasso, Biancavilla, Bronte, castiglione di Sicilia, Giarre, Linguaglossa, Maletto, Macsali, Milo, Nicolosi, Pedara, Piedimenote Etneo, Randazzo, Ragalna, S. Alfio, S. Maria Di Licodia, Trecastagni, viagrande and Zafferana Etnea.

As noted earlier, transportation infrastructure is regarded as one of the most important trigger of change that is detectable and extractable from RS in term of development of cultural tourism.

Hence, this aspect of the urban landscape needs to be monitored for planning purposes, updating cartography, and as landmarks for automated navigation and experiment databases (Karimi *et al.*, 2003 and Vanschoren, J., *et al.* 2012).

New paths or greenways can potentially be considered the first sign of cultural itineraries in these sectors.

This can be regarded as a move from zone C to zone B in the operational classification of the greenway network change. With this information, prevention of environmental deterioration can be facilitated through improved and more targeted planning policy related to development at the critical areas and particularly devoted to various disastrous human interventions (such as flow variations, excavation works and dike building). Furthermore, as greenway networks become denser over time, the conversion of adjacent land uses and new potential greenway routes can be identified and analysed with satellite imagery in a timely and efficient manner.

Extracting greenways and cultural building infrastructure from satellite imagery is dependent on image resolution. However, what can be extracted is useful in terms of indicating where development is occurring and, in particular, identifying the density of this development.

Most objects, including buildings, are complex with respect to both their spectral and spatial characteristics, making them somewhat difficult to detect and classify (Zhang, 1999 and Rozenstein, O. Karnieli, A. 2011).

Since buildings are one of the most important classes in land use and in the identification of cultural and environmental itineraries, their identification and classification is important in terms of cultural tourism. As the densification of buildings occurs, in terms of land use classification, a movement from zone D to zone C is likely occurring (Figure 2-4). Through the determination of building location and survey, the enhanced conservation and protection of natural and cultural environments can occur and the provision of required environmental infrastructure can be achieved. Depending on the rate of densification of buildings, there exists an increasing potential for environmental stress as buildings are erected relatively quickly and without adherence to proper zoning by-laws and construction quality guidelines.

2.8 Ground-based GIS Indicators

The ground-based GIS indicators used in this research, shown to the right in Figure 2-4, are all derivable from land use-based data and the "Carta della Natura" able to reply both to management need of related protect area human activities (release of environmental controls) typically collected by local and national public institutions and in particular the Etna Park Institution.

However, without a properly designed and implemented framework to analyse the data, their potential is unexploited. The main indicators in this research that can help to identify locations of potential environmental values and to limit the effect of stress, and that are derived from ground-based GIS data, include:

1) acreage of land use / land cover under rural vs. urban use;

2) population, population density;

3) quality of vegetation and distribution of protected species;

4) environmental infrastructures.

These indicators are not used independent of one another. Their individual and combined impacts point to potential environmental quality and to limit the effect that may be occurring at the zoning as urban growth occurs. Furthermore, these ground-based GIS indicators are analysed via an integrated process with indicators derived from RS imagery.

The integration of indicators derived from RS imagery and ground-based data and land use analysis within a GIS provide further spatial analysis capabilities beyond what either can reveal alone. Each of these ground-based indicators is now discussed in greater detail.

First, acreage of land use by land use class provides the same type of information as the satellite imagery analysis and can be used as collaborative data to validate the imagery, assuming a similar time frame. Similar to the RS derived land use indicators discussed in the previous section, in terms of the operational classification of land use, the greater the acreage of land still under agricultural use, the more likely that this land would fall into zone C and D (Figure 2-3). As noted earlier, it is the rate of change that determines the relative potential for environmental stress evidenced by the amount of change in the indicator under examination.

This type of scattered development, often referred to as "leapfrog development" because of the dispersed and random growth, has been criticized for its inefficient use of land resources and energy (Yeh and Li, 2001).

Population, population density, numbers of dwellings and migration levels are all indicators of growth and sprawl. The difficulty with these indicators is that ground-based population measurements and estimates are not always very accurate (Cowen et al. 2000). Even when a land use nalysis has been completed, the results often take a long time to process and publish, and the quality and validity of the data are questionable. However, with these constraints noted, the data that are collected and reported can be used to assist in the management of the urban living and working environment. With this information, SEM can be better facilitated because without such data, there is little indication of where growth may occur and how quickly the landscape is likely to change from rural to urban uses. The operational use of indicators derived from GIS-based data analysis to classify urban fringe land use can, in this context, be successful in terms of identifying the location of this land use change by accurately identifying potential stress areas.

Finally, the quality of vegetation and green infrastructures are surrogate measures for CT. For SEM and CT to occur, transportation must be observable and quantified in terms of growth and density of greenway network patterns.

Just as the extraction of greenway from satellite imagery can be used as the skeleton around which development occurs, the quality of vegetation registered can be used as supplementary data to help confirm the findings from the satellite images. In zone C and D of figure 2-3, minimal greenway development is evidenced. However, as land use evolves into zone D, a detectable characteristic is the development of the greenway network.

The ability to identify greenways in contrast to other forms of growth and in accordance with environmental values is uniquely important in terms of zoning classification and in the context of this research. Specifically, greenways must be identifiable as they form the main trigger of development for CT. The indicators outlined above that are derived from RS imagery and GIS-based ground data analysis can be methodically combined to classify land use.

The operational use of SIT based on these indicators (Figure 2-3) is integral to the staging of development occurring at the zoning in terms of accurately identifying areas of rapid land use change in order to detect new environmental values and cultural itineraries and to limit the effect of potential environmental stress. This classification and identification process is fundamental to the goals of SEM in terms of their successful application to

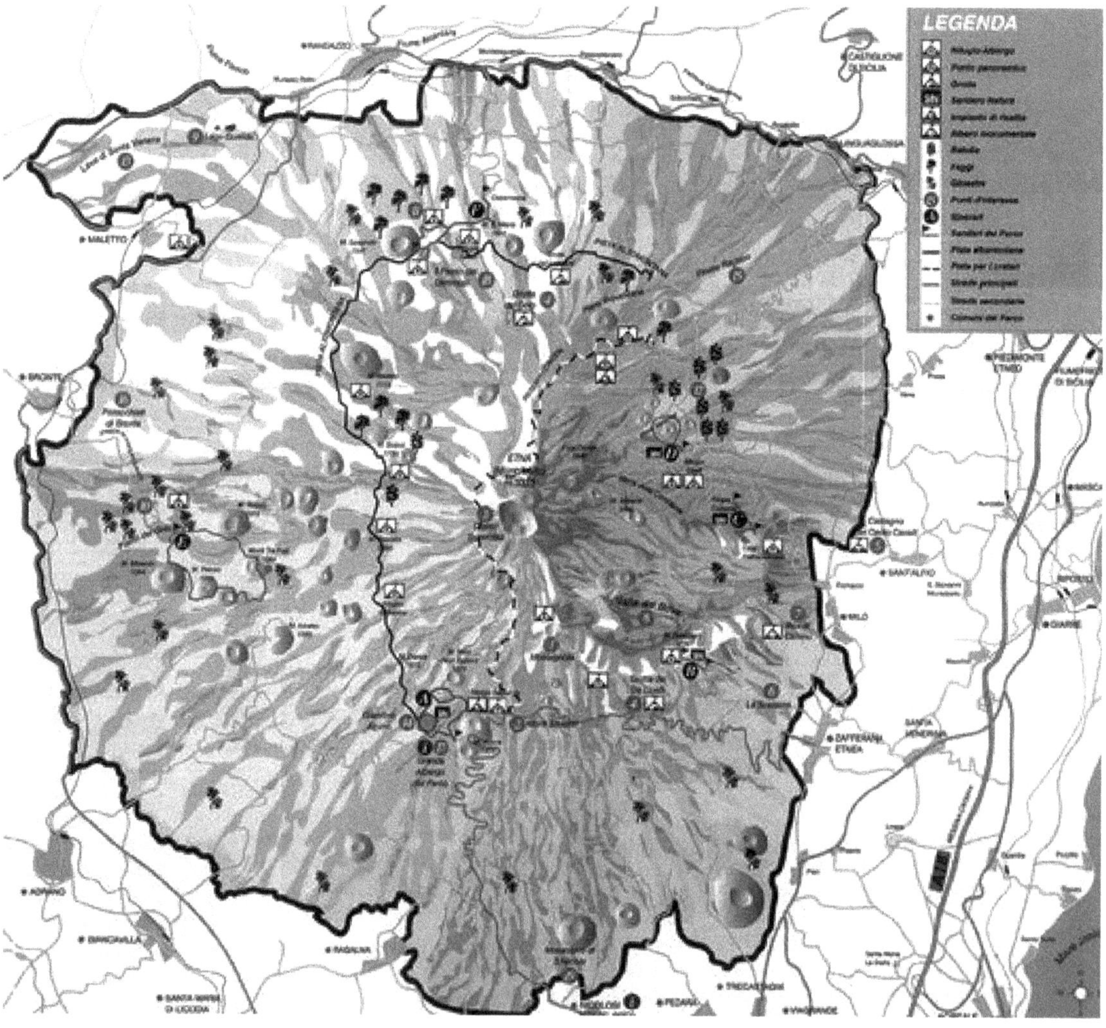

FIGURE 2-3: THE INTERACTIVE LAYERS AND MAPS OF THE GIS (PRODUCED IN COLLABORATION WITH THE GISAT WWW.GISAT.IT AND FINANCED BY P.O.R. 2000-2006 - 1.11 "INTEGRATED INFORMATION SYSTEMS"WWW.PARCOETNA.CT.IT/SITGIS.HTM)

planning guidelines and policy creation in Park Institutions and for community decision making, not only for the Etna regional Park (Ghose, R. 2011).

2.9 Greenway Network and Cultural Sources: theory and research

"Greenway are a major element of the environment. They define the circulation pattern and the geometrical configuration of the environment. They are often the beginning of neighbourhood interaction connecting individual dwelling units and serving as a `stage' where people interact with bits of the total environment" (Adepeju, A. and Oluwole, O. 2012).

O'Sullivan et al. (2000) stated that mobility in natural environment, defined as the ability of people to travel over nature, increases due to the necessity of people to move to and from architecture, places of natural and anthropic sources, and leisure facilities. Further, he noted that as these localities become more geographically dispersed, the impacts on and requirements of, the existing green infrastructure network increases, precipitating growth in the network. There exists a necessity for being able to locate and map the route networks in a more timely and efficient manner. This not only includes existing routes, but must also include the need for better planning of new cultural routes as well as the need to detect and monitor new greenways as they develop (O'Sullivan et al, 2000). This temporal examination of old and new routes is necessary to facilitate the analysis of land use change over time as secondary green infrastructure grows around the developing transportation network. As noted above, these needs form the fundamental basis for this research. Roads have traditionally been taken for granted due to their economic and social significance as transportation and utility corridors. They can also have a significant impact on the environment by transversing areas of wildlife habitat, and are one of the most destructive elements in the habitat fragmentation process.

Roads have traditionally been taken for granted due to their economic and social significance as transportation and utility corridors. They can also have a significant impact on the environment by transversing areas of wildlife habitat, and are one of the most destructive elements in the habitat fragmentation process (Viles RL et. alt. 2001). This research also analyses the degree to which the 'greenway' and 'connectivity' concepts can be applied in an integrated approach to managing road reserves and the adjacent landscape to facilitate the extension of ecological corridors and mitigate adverse environmental effects of the roading network. While the Etna landscape and road networks are currently managed independently, the study found that there is potential opportunity to use the Etna road network as greenways to achieve partial integration between conservation and land use systems. A coordinated approach between conservation managers, road construction and management authorities, local government, utility managers, landowners and others who have interests in roadside management is necessary if integration is to be successful. Changes are also needed in the design of roads within reserve areas so that road safety requirements are augmented by consideration of the character of the landscape matrix and the needs of species moving through the landscape.

The ecological corridors or greenway are natural areas able to connect zones with peculiar naturalistic-environmental valences" (ECNC, 1996). The 'greenway' and 'connectivity' concepts can be applied in an integrated approach to managing road reserves and the adjacent landscape to facilitate the extension of ecological corridors and mitigate adverse environmental effects of the roading network.

Mautone M (2001) found that there is potential opportunity to use the road network as greenways to achieve partial integration between environmental conservation and land use developement. A coordinated approach between conservation managers, road construction and management authorities, local government, utility managers, landowners and others who have interests in roadside management is necessary if integration is to be successful. Changes are also needed in the design of the greenway network within reserve areas so that road safety requirements are augmented by consideration of the character of the landscape matrix and the needs of species moving through the landscape. On the other hand, as land use changes, wildlife habitat and migration corridors are lost and normal ecosystem functions are disturbed or destroyed. Using GIS and principals of landscape ecology, Weber T., Wolf J. (2004) at the Maryland Department of Natural resources identified an interconnected network of "hubs" and "corridors" that were the focus of state and local agency deliberations and revisions. Elements of the network are being prioritised for conservation and restoration based on ecological parameters (e.g. sensitive species, important wetlands or riparian zones, etc.) and threat parameters (e.g. protection status, development pressure, etc.). The goal of the GIS under the Green Infrastructure Assessment (GIA) programme was to help identify and ecologically sound open space network, and ultimately, to incorporate the agreed upon network into state and local land conservation planning.

Hoctor TS, Carr MH, Zwick PD (2000) used geographic information systems software (Arc-Info) to develop a decision support model that uses land-use data and information on significant ecological areas-including important habitats for target species, priority ecological communities, wetlands, roadless areas, floodplains, and important aquatic systems-to identify larger areas of ecological priority and potential ecological linkages. The result of this process, the Florida Ecological Network, includes approximately half the state's area, with over half of this network already in conservation lands or public-domain.

There exists a necessity for being able to locate and map the route networks in a more timely and efficient manner. This not only includes existing routes, but must also include the need for better planning of new routes as well as the need to detect and monitor new greenways as they develop (O' Sullivan, 2000). This temporal examination of old and new routes is necessary to facilitate the analysis of land use change over time as secondary infrastructure grows around the developing transportation network.

As noted above, these needs form the fundamental basis for this research.

These analysis are then added to a multi-source GIS and experiment database for analysis in conjunction with other indicators: environmental sources, cultural heritage sources, etc (Vanschoren, J., et al. 2012).

We should add that "Heritage" is now taken to mean not only prestigious monuments but also major rural or industrial buildings and sites, as well as the less tangible heritage of popular traditions. This new concept of heritage was eventually actuated with the national law n. 137of the July 6, 2002. This Decree, called "*Codice dei beni culturali e del paesaggio*" (art. 10 n. 137), sets down the regulations to achieve the correct balance in the conservation and the valorization of the cultural heritage.

The word "heritage" in its broader meaning is generally associated with the word "inheritance." that is, something transferred from one generation to another. Owing to its role as a carrier of historical values from the past, heritage is viewed as part of the cultural tradition of a society. (Nuryanti, 2004).

In this field, from 1969 to 2000, CIPA was the Comité International de la Photogrammétrie Archi-tecturale, jointly founded as an ICOMOS Scientific Committee under a contract with ISPRS, the International Society of Photogrammetry, Remote Sensing and Geoinformation. CIPA decided to generalise in 2012. This is a trend among many working groups, in connection with the progressive generalisation of computing and information technology. Since this decision, CIPA is known as the International Scientific ICOMOS and ISPRS Committee on Documentation of Cultural Heritage, and is working towards the following topics: 1. General Recording, Documentation and Information Management (RecorDIM) ICOMOS Tools 2. Conservation Management, Research, Analysis, Design, Maintenance and Monitoring Tools for conservation specialists involved in:

- Archaeological Heritage Conservation
- Architectural Heritage Conservation
- Cultural Landscapes Conservation
- Conservation Project Management
- other fields of conservation.

Tools for heritage recording and documentation are the following ones:

- Photographic Recording and Documentation;
- Close Range Photogrammetry and 3D-Scanning;
- Surveying and Mapping;
- Aerial Photogrammetry for Cultural and Natural Landscapes;
- Geophysical and Aerial Methods for Archaeological Prospection;
- Underwater Technology and other special methods of surveying, mapping and data collection;
- Information Technology at large for all kinds of Heritage Information Systems.

CIPA is a partner in the RecorDIM Initiative that has the goal to 'bridge-the-gap' between Information Users and Information Providers. This Initiative's activities are outlined in a RecorDIM status report available at: http://cipa.icomos.org/reports.html. CIPA invites all National and International Committees of ICOMOS and all experts in the above areas to co-operate in CIPA's Working and Task Groups, and to nominate National and Committee Delegates as Liaison.

On other important source is the geographical database of cartographical data and satellite and aerial photography images of Italy, developed by CNR and by the Etna Park Institution. It can be used as a basis for a census of the Italian artistic heritage and of artistic sites to be protected if threatened by a catastrophic event. The geographical-cartographical data available *in situ* is composed by:

1. Digital Elevation Model (dem), at a gauge of 240m and 70m, of the whole of the Etna Park.

2. Landsat satellite images (Landsat TM)

3. SPOT Multispectral and Panchromatic Images

4. 2001 MIVIS Images (by the CGP for the INGV)

5. Digital cartography 1:25.000, 1:50.000, 1:100.000, 1:250.000 taken from the maps of the Istituto Geografico Militare (Military Geographical Institute).

6. Orthophotographs of Italy (1997)

7. Administrative boundaries (regions, provinces, municipal districts) and related data (eg. number of inhabitants per municipal district, etc.). Source: istat.

All the images are available in different reference systems (Gauss- Boaga and UTM) and constitute a continuum across Italy. It is therefore possible to move virtually across the national territory, with different zoom views, observing different databases, while maintaining continuity and preservation of distance.

In particular within the context of the "Progetto Finalizzato Beni Culturali" of the CNR, the GIS has been used for a census of the archaeological sites of the area of Vesuvius and Etna. Three different zones of volcanic hazard can be identified: red zone, affected by dangerous proximal phenomena (high temperature pyroclastic flows, ballistic fallout, mudflows, pressure waves, etc.); yellow zone, affected by ash fallout and the blue zone, affected by flooding/mudflows. Geographic Information Systems (GIS) are an ideal means to manage data of cultural landscapes or cultural heritage objects in landscapes (Heinz 2004). Quite a few World Heritage landscapes and objects are already documented (and managed) using GIS-techniques (Roessler 2000). Special guidelines for the use of GIS in cultural resources management were developed (UNESCO 1999).

Considering the size and complexity of a landscape, adequate tools have to be provided for its management. After UNESCO had introduced legal instruments to recognize and protect cultural landscapes of outstanding universal value, and several cultural landscapes were included in the World Heritage list (http://whc.unesco.org/), it was soon realized, that "the resources management, e.g. the management and protection of the cultural values for which these sites have been inscribed, has to be improved" (Roessler 2000). It is obvious that surveying and mapping procedures, based on sufficiently complete and updated databases, are needed to document and monitor the sites and to allow predictive modelling of future developments. The role of Geographic Information Systems is of special importance in this context (UNESCO 1999). A GIS provides tools to store, manage, analyze and visualize spatially related data. If a GIS with a temporal dimension is chosen, the evolution of a cultural landscape can be modelled and studied.

In general, 'Green infrastructure' is a term used to describe systems of parks, greenways, open spaces and other natural landscape elements that provide community benefits. Although we have some understanding of how people use parks and developed greenways, little has been documented about use of the undesignated public and private spaces along green infrastructure features such as stream corridors. The purpose of this research was to examine characteristics that may influence people's use of undesignated open spaces along the stream corridors that form the skeleton of many green infrastructure systems. Data were obtained from a Recreational Use Attainability Analysis (RUAA), an evaluation performed for the Italian Commission on Environmental Quality.

Results indicate that the proximity of stream corridors to local residents, the level of pedestrian access available and tree cover were the best predictors of recreational use. Observations also indicated that urban and suburban stream corridors afford a variety of recreational and aesthetic values to residents. Implications for policy, planning and design of green infrastructure are discussed (Shafer, C.S., Scott, D., Baker, J. & Winemiller, K., 2013)

2.10 Automated extraction of man-made and natural objects

Other researchers have examined the possibilities of extraction and classification of man-made features in environments including, for example, cultural buildings and greenway network (Trinder, 1995; Haala and Brenner, 1999; Shettigara *et al*., 2000; Sowmya and Trinder, 2000; Guindon, 2000; and Baltsavias, 2000). Shettigara *et al*. (2000) defined the terms 'artificial' and 'man-made' as objects that have been artificially created using synthetic or natural materials.

Automated extraction of man-made and natural objects from aerial photography and satellite imagery is one of the fundamental tasks in image analysis (Kim *et. al,* 2001 and Mahesh Pal, M. et al 2013). Attempts to extract specific features automatically from digital images have been in practice for more than fifteen years using many different methods and algorithms (Gruen, et al 2003 and Barranquero, J., *et. al* 2013). The desire (and need) to extract linear features has been of interest in the private sector, government and the military. A basic objective of this research is to establish an operational process of linear (greenways) feature identification from RS source imagery and subsequently to extract these features into a GIS. The term 'extraction' itself can be vague at times (Doucette et al., 2001). Here, the task of greenway extraction addresses the issues of both identification and subsequent delineation.

Before any extraction and/or classification of such features from RS imagery can take place, the geographic context of the images must be understood. Context, in this sense, can be defined to include, using Couclelis (1999) terms, the site and situational dimensions, as well as the spectral and spatial resolutions of the source imagery.

However, this can be an important consideration in terms of representing inherent or local knowledge in the context of RS images. Factors such as the size of the object in question with respect to the image pixels under examination influence the ability to extract and classify man-made features from other information contained in the image. Even after understanding and accounting for these factors, specific feature extraction has traditionally been completed only via semi-automated methods (Sowmya and Trinder, 2000 and Mahesh Pal, *et. al* 2013). Hence, recent attempts have been made by researchers at automating the feature extraction process.

Methods used include object recognition models using spatial/spectral/contextual attribute combinations in high resolution imagery for recognition of such features as greenways and cultural heritage (Guidon, 2000); digital ortho-photos through 3-D extraction from stereo pairs (Baltsavias, 2000); finding edges of man-made objects and extracting them by mapping the boundaries between the man-made and adjacent natural objects (Shettigara *et al.*, 2000); and using laser scanners to survey, classify and extract buildings from environments (Haala and Brenner, 1999) and to survey the green infrastructure (Shafer, C.S., et. al 2013).

This availability of survey approaches gives the idea to use them for the territorial planning based on buildings listing, following the ICCD (Central Institurte for the Listing and Documentation) standards, on architectural inventory, on cultural landscape and heritage georeferencing. The success of these methods has been varied. However, each has achieved the extraction of features via either manual or semi-automated techniques.

As more methods are developed for data acquisition via successful identification and extraction of features for digital mapping and GIS, knowledge representation modelling skills are used to adapt methods to particular characteristics of the image (Sowmya and Trinder, 2000 and Yıldız, O. T. 2011).

In particular, the extraction of linear features has received a great deal of attention and, subsequently, different methods of route extraction have been developed and tested.

These include, but are not limited to, map-image modelling, edge detection and filtering, dynamic programming. These techniques are described in the following capters.

Land development and hence land use change from one state to another, typically occurring around existing built-up areas and along routes (Yeh and Li, 2001 and Yıldız, O. T. 2011). As these changes are invoked, they become the triggers for further development through a cyclical process that is both temporally and spatially variable. As greenways appear, typical landscapes develop naturally along and adjacent to the green infrastructure. Consequently, this feature represents the catalyst for new development.

As noted above, the analysis of satellite imagery for linear feature extraction has been extensively researched (Mayer and Steger, 1998; Steger *et al.*, 1995); Fiset *et al.*, 1998; Jedynak and Roze, 1995; and Barzohar and Cooper, 1995). Modeling methods such as abstraction, defined by Mayer and Steger (1998) as the increase in the degree of simplification and emphasis, are used to detect relatively homogeneous regions in an image. In the case of greenways, their surface and reflectance is not necessarily homogeneous and of a different brightness than the surrounding areas. Using this assumption, Steger *et al.* (1995) extracted a partial route network. However, shadows present in the image produced erroneous greenways segments in some areas.

Fiset *et al.* (1998) modelled map-image matching for route network updating. Because the updating of topographic maps is traditionally a tedious and time-consuming task, they sought to automate the process. Their process began with the rasterization of the existing route network. The rasterized network map was then matched to a classified image to locate potential new greenways that could be derived from the image.

Edge detection and filtering have also been used for linear feature extraction (Tripathi and Gokhale, 2000; Karimi *et al.*, 2003; and Vosselman and de Knecht, 2002). The process of greenway feature extraction consists of three main tasks, including image acquisition, feature extraction, and input to database management systems. Seeded segments were processed with the semi-authomatic approach to locate and extract turistical routes and features in a semiautomatic manner. This research using high-resolution imagery showed promise as a means of road feature extraction. However, before it could be fully automated and implemented effectively, several aspects needed to be refined. These included:

1) shadows cast by buildings, trees, etc.;

2) barriers like bridges and overpasses;

3) occlusion due to vegetation;

4) confusion between greenways and other man-made construction;

5) differences in greenway materials used;

6) traffic pattern differences;

7) position errors caused by elevation change; and

8) limitation of the information content on the images.

(Karimi *et al.*, 2003)

They cause problems and inconsistencies for all methods of feature identification and extraction and need to be researched further before fully automated processes can be realized. Because the extraction and classification of detailed greenways features from satellite imagery for input into a database management systems, especially a GIS, has not been extensively reported in the literature, it represents a gap in the research and an area that deserves further exploration (Karimi *et al,* 2003, Cavalli *et al,* 2005, and Yıldız, O. T. 2011).

Since fully automated methods, with no human assistance, for conversion of remotely sensed images to digital maps are still generally unfeasible, semi-automatic methods that involve computer and human interaction are the most viable alternative (Vosselman and de Knecht, 2002). In this context, greenway identification by filtering can be used to estimate the parameters that describe a route by shape and position. Vosselman and de Knecht (2002) used a Kalman filter to detect and remove outliers in observed image feature characteristics to facilitate continuous feature mapping. They described greenway feature characteristics as typically being represented by the following characteristics: are elongated; their surface is eterogeneous and sometimes has a good contrast with the adjacent areas (Vosselman and de Knecht, 2002). Those greenway feature characteristics are not sufficient by classify the greenway network.

European Centrer for Nature Conservation (ECNC), reviews a previous work concerning the research of ecological corridors between the Vesuvian area and other neighbouring natural sites (Parthenius, Picentini and peninsula Sorrentin). The work was based on the August 2001 image acquired by the Aster sensor aboard NASA 's Terra satellite (from 15m to 90m spatial resolution, 14 bands spectral resolution). The application of a vegetation index and of a textural analysis (based on green band of the sensor) has been carried out on the referring image, with the purpose of identifying several surfaces far the development of ecological corridors or greenway or natural areas able to connect zones with peculiar naturalistic-environmental valences (ECNC).

Semi-automatic extraction of green infrastructure networks using dynamic programming has been researched extensively by Gruen *et. al.* (2003, 2004) and by GFOSS (2013, www.gfoss.it). Their research defined a typical route extraction procedure as consisting of several steps including route sharpening, route finding, route tracking, and route linking in orted to detect and extract routes from single SPOT scenes and digital aerial photographs. Similarly to Vosselman and de Knecht (2002), they listed several generic properties of features that were used for identification within the image. These features included the properties that a road pixel is lighter than its neighbours on both road sides, grey values along a road typically do not change within short distances, and a road is linear, smooth, and of generally consistent width. Dynamic programming was defined as a technique for solving optimization problems (Gruen and Li, 2002) and Bachmann, C., (2013) as it can simultaneously evaluate unrelated road characteristics such as those outlined above. Optimization, in their context, was defined as a determination of the best multistage process for road extraction. By interactively inserting vertices into observable road segments in the image prior to processing, these seed points were then connected to form polygons. Results show that dynamic programming was successful, particularly in rural areas where the distinction between greenways and the surrounding agricultural land use was very clear. Use of Least Squares B-spline (LSB) Snakes is a further extension of the work done by Gruen and Li (2002) and by GFOSS (2013, www.gfoss.it). LSB-Snakes represent an approximation of the position and shape of a linear feature within a RS image or scanned aerial photograph. The methodology combined the tools of ordinary least-squares estimation with the determination of energyminimizing functions. Using this approach, Gruen and Li (2002) were able to use only a few coarsely distributed seed points set by a human operator near the features of interest to extract features using least-squares regression. In their tests, the process was able to extract greenways under difficult conditions including varying widths and obstruction by buildings, trees, and vegetation. Barranquero, *et al.* (2013) concluded that the results clearly prove the success of the selected algorithm.

A different method of extraction, which dates back to the late 1980's, is the use of computational neural networks for the classification of remotely sensed imagery (Wilkinson, 2004 and GFOSS researches). Since the potential of neural networks was first demonstrated in the context of feature extraction, there has been a quickly growing body of literature on this topic (see, for example, Benediktsson and Sveinsson 1997, Liu *et al.* 2000 and GFOSS). In recent years, the number of successful applications of neural networks for image classification has been increasing. Liu *et al.* (2000) and Ciresan, D. C., *et. al* (2012) claimed that neural network classifications were more accurate than conventional approaches to remote sensing for several reasons:

1) neural network classifiers make no priori assumptions about data distributions (and they) are able to learn patterns in the distribution classes;

2) neural networks can readily accommodate collateral data such as textual information, slope, aspect, elevation; and

3) neural networks are quite flexible and can be adapted to improve performance for particular problems.

Following from this, Liu *et al.* produced in MATLAB software several programs for the visualization and analysis of network infrastructure. Similarly, Bendiktsson and Sveinsson (1997), used computational neural networks as an alternative to conventional classifiers for the classification of remote sensing data. Given the successes generated in the use of neural networks, it is likely that more implementations in this field will appear in the future (Ciresan, D. C., *et. al* 2012).

However, in terms of their commercial implementation, packages such as PCI's EASI/PACE Imaging Kit are just now including back-propagation neural network algorithms. Because of the expense of this package and the lack of knowledge on its use, a large gap currently exists, and will likely continue to exist, between research and software implementation (Liu *et al.*, 2000 and GFOSS).

Alternatively, laser scanning is a relatively new (in this context) and independent technology, which has made a recent resurgence in the study of image feature extraction (Ackermann, 1999; Haala and Brenner, 1999; and Wehr and Lohr, 1999 and Ciresan, D. C., *et. al* (2012). Generally, laser scanning is captured on airborne platforms rather than earth-orbiting satellites.

However, the techniques of data gathering to measure the visible ground surface or objects on it make it useful for linear feature extraction. Ackermann (1999) pointed to laser scanning as a particularly interesting new application for the automatic capture of buildings and built-up urban areas within digital urban models. To this end, Haala and Brenner (1999) used airborne laser scanning to classify urban scenes and complete a subsequent process of building reconstruction into a 3-D visualization. Because the use of laser data proved to be successful for them in reconstructing urban environments, they suggested that airborne laser scanning was useful for the automatic generation of urban databases.

Finally, reconstruction using 3-D edge extraction has been developed by Gruen and other collaborating researchers (Zhang *et al.*, 2000; and Zhang and Baltsavias, 2000 and GFOSS). The continued goals of researchers is to build an automatic and robust system to reconstruct route and networks from aerial images with the aid of existing several road databases and other data (Zhang and Baltsavias, 2000 and Shafer, C.S. 2013). In this case, they developed a new approach for automatic road network extraction by integrating known techniques for processing imagery with existing spatial databases. Their existing data for networks included road type, class, width, length, topology information, and land cover. These ground-based data were then integrated with the imagery for 3-D road extraction. Zhang and Baltsavias (2000) tested several combinations of this integration and were able to extract successfully a road from a suburban setting. Zhang *et al.* (2000) found that the 3-D edge extraction method provided reliable results given the large amount of existing knowledge input into the process. Based on a visual interpretation of the imagery provided in Zhang *et al.* (2000), it would appear that this process, like previous research into network extraction, has had far greater success in rural areas as compared to the more diverse and complex urban scenes and road networks (Shafer, C.S., 2013).

The feature extraction tools reviewed above all have their strengths in terms of adequately extracting linear features for digital map creation and updating. In particular, automatic extraction methods have been investigated for the building reconstruction. It is found that buildings with closed polygon footprints are difficult to obtain the desired accuracy. In this research, a semi-automatic approach is proposed for the extraction of closed polygon footprints from lidar data. Two types of closed polygons have been investigated, where one type is a simple closed polygon and the other is a complex combination of multiple polygons. From the reference data that is surveyed with a total station, the building footprints have been extracted. The extracted footprints from lidar data have been compared with the surveyed footprints using a confusion matrix and Kappa Index. It is concluded that the approach is reliable because the closed polygon footprints can be extracted with a high Kappa Index. This indicates that the proposed procedure is suitable to be utilized for cadastral mapping of remote built up areas (Shirowzhan, S. and Lim, S. , 2012)

Much more research is required in this area before this type of stand-alone method becomes routinely used for urban planning and land use management purposes. This research does not seek to improve upon the technical process of skeletal linear feature extraction. Rather, the integration of the extracted features with ancillary data for identification and classification of areas of potential environmental value is the objective. Once greenways are successfully identified and extracted from RS source data, the question of how these data can be combined with other data sources to locate areas of potential environmental value is addressed. While the greenway network can be viewed as the skeleton around which cultural heritage conservation and valorization occurs, integration of this information with other indicators of change in a GIS (such as land cover classification and data) allows for the pattern and process of land use change to be monitored and better understood. Thus, the roads are regarded as the integral indicator in terms of detecting and locating potential environmental values and stress in a temporal analysis process. The greenway mapping also represents a key issue for environmental planning and conservation and for defining monitoring programs.

2.7 Summary

This chapter has summarized the current situation in park institution in terms of the application of the concepts of CT and SEM for the conservation and valorization of cultural sources. Combined, these define the concept of SEM, which is the desire for sustainable urban growth and management in conjunction with cultural heritage development strategies. SEM was then placed in the context of using RS analysis and SIT to achieve SEM and CT. Specifically, the discussion focused on the concept of urban environmental stress and the use of SIT to detect and analyze potential stress triggers. Throughout the discussion it was noted that without access to spatial information, it is difficult to develop and implement growth management strategies, especially at the zoning level. To this end, an integrated framework model was presented that linked the concepts of CT, and SEM with the use of SIT to facilitate potential stress detection analysis in one hand and to valorise the cultural sources on the other.

The current state of research on the integration of RS and GIS for the detection of environmental stresses was also reviewed. In particular, the discussion focussed on the detection and extraction of features that are possible indicators of environmental stress on one hand and the analysis of environmental network from satellite imagery, their union with ground-based indicators, and their subsequent integration to facilitate SEM and CT. To achieve this detection, a model was presented to operationalize the integrated framework model in order to facilitate land use classification.

In the context of this project, the Catania Province and its hinterland is the case study for the operational implementation of the integrated framework outlined in this chapter. As such, Catania Province's past and current planning practices are discussed further in the following chapter along with a detailed review of the research design implemented in this project.

CHAPTER 3
RESEARCH DESIGN

This chapter discusses the methods and operational design of the research undertaken in this research. To set the stage for the study area discussion, the past and current planning frameworks of cites in the Etna Regional Park, and specifically those of Catania Province, are discussed. Next, the process of selecting the study area, field research experiences, the specific research methodologies used, data collection, data processing and analysis methods are all discussed in sequence. Each aspect of the research design relates directly to the components of the integrated framework presented and discussed in Chapter 2.

3.1 Study Area: an example of test sites

"I wonder where Etna is. The atmosphere below us was blanketed with clouds and this effect produced a wonderful phenomenon above our hands. This illusion was streaked with white and grey and seemed to be some bodily thing, but how can a body float in the sky. The driver explained to us that our admiration was drawn to a flank of Mt. Etna that appeared between breaks in clouds, snow and mounds creating the streaks, and that it was not the summit yet". (Wolfgang Goethe, 1787)

Goethe's first impressions of Mt. Etna, from his trip there in 1787 are still fundamentally accurate after hundreds of years later. The Mt. Etna's ability to capture an entire spectrum of natural and cultural environments in a single park is part of the reason Mt. Etna (or "Mons-Gebel") was selected as the study area for this research (fig. 3.1).

FIGURE 3-1: ETNA PARK AND CATANIA PROVINCE (AS AN EXAMPLE OF TEST SITES

The presidential Decree of the Sicilian Region n. 37 of March 17, 1987 established the Etna Regional Park and laid down its boundaries and activities permitted in each sector. The regional law n. 89 of May 6, 1981 and n. 14 of August 9, 1988 set down the regulations for the establishment of natural parks and reserves in Sicily.

Mount Etna is inscribed on UNESCO World Heritage List.

With a population of over 200.000, Etna has experienced an average annual population growth rate of 0.05 percent over the last ten years (ISTAT 2001). In addition to this, it had the distinction of being the European's highest volcano. It is characterised by an area of 59.000 hectares and 20 small towns in its hinterland.

The eruptive history of the volcano can be traced back 500,000 years and at least 2,700 years of this activity has been documented. The almost continuous eruptive activity of Mount Etna continues to influence volcanology, geophysics and other Earth science disciplines. The volcano also supports important terrestrial ecosystems including endemic flora and fauna and its activity makes it a natural laboratory for the study of ecological and biological processes. The diverse and accessible range of volcanic features such as summit craters, cinder cones, lava flows and the Valle de Bove depression have made the site a prime destination for research and education.

In particular the Catania Province has his boundaries in Santa Maria di Licodia in the Weast direction and Giarre in the in the Nord-Est direction of the "Metropolitan area", with a population of over 733.833 (ISTAT 2001). It is characterised by an area of 952,11 Kq. and 27 small districts in its hinterland.

Contrary to common belief, Etna is not a simple shield or strato volcano. Its volume is about 500 km^3 with a central summit area of 1.600 km^2. The bigger axes is about 600 km in the Nord-Sud direction and the Est-Ovest axes of about 150-160 km. (Pesaresi, 2003). Its shape and structure are extremely asymmetric and complex, and a classification of the mountain on a morphological basis is nearly impossible. The reason for this complexity is that Etna did not grow as one single large cone, but as a succession of volcanic edifices most of which suffered partial collapse at least once during their lifetimes, and whose centres shifted from one place to another. Several calderas can still be discerned morphologically, the most impressive being the huge Valle del Bove (Valley of the oxen) on the eastern side of the volcano whose latest collapse episodes occurred only a few thousand years ago. Etna's morphology is further complicated by the presence of areas with more frequent eruptive activity, like the Northeast Rift which forms a distinct ridge dotted with eruptive cones and cut by numerous gaping fissures. Hundreds of minor pyroclastic cones are scattered all over the flanks of the mountain, some of which appear to be small volcanoes on their own, with edifice heights of several hundred meters, while many others are much smaller, with heights of only a few tens of meters. These were produced by flank eruptions, the most dangerous, and probably the best documented, type of activity at Etna in recent times. At the summit of the volcano stands a complex of large cones which actually host the four summit craters. This peculiar family of craters, which are somehow connected but show a surprising degree of independence from each other, is a relatively recent feature. One hundred years ago, until 1911, there was one single large cone at the summit of Etna, that was truncated by the 500 m-diameter Central Crater. Two new craters, the Northeast and Southeast Craters, formed in 1911 and 1971, respectively, and have since built their own cones which rival the old central summit cone in size and height. In recent years the Southeast Crater has been particularly active and its growing cone now forms a prominent landmark at the summit of Etna. Viewed from south and southeast it actually seems higher than the central summit cone, but this is an effect of perspective; actually the summit of the Southeast Crater cone is still about 20 m lower than the highest point of the volcano.

3.1.1. Dike buildings in specific areas and zoning

The population growth rate has two primary spatial manifestations, one internal (in particular zone A) and the other at the periphery of the Etna Park, in zone C and D (Lo Tauro et alt. 2005)

In figure 3-2 we analised process of listing of dike buildings.

The former has the effect of increasing already high various disastrous human interventions (such as flow variations, excavation works and dike building) in specific areas. The latter dimension of growth has had the effect of pushing development out into hinterland areas of productive agriculture or idle land use. This growth has implications for loss of agricultural productivity and extending urban services further from traditional service hubs; often to the point where new, dispersed service hubs have evolved. The current situation resulted in additional uncontrolled urban growth and dike buildings particularly concentrated in the zone D (Fig. 3.3).

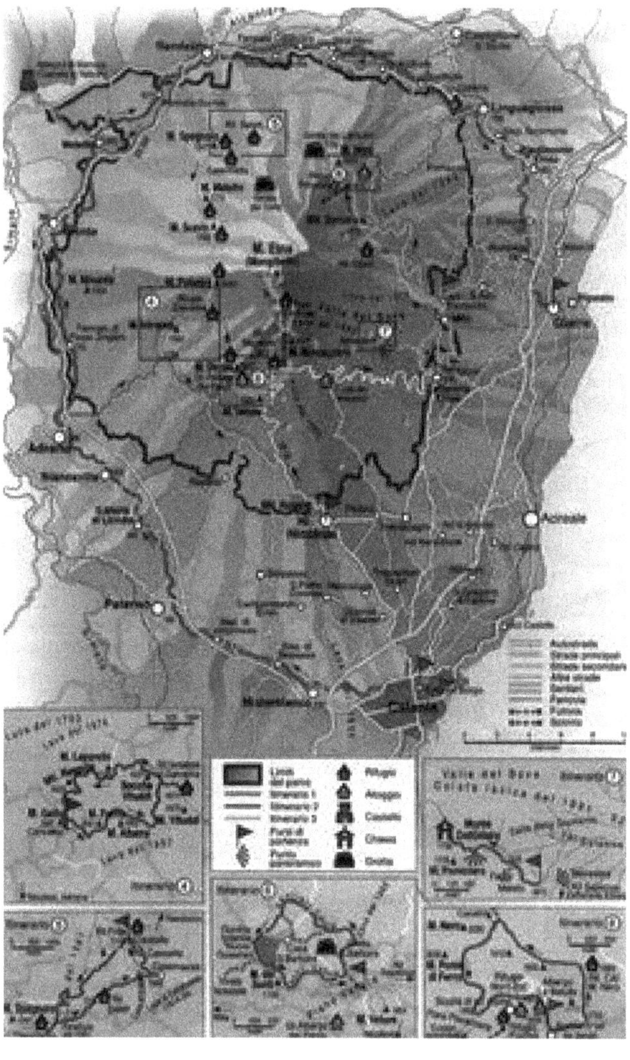

FIGURE 3-2: LISTING OF DIKE BUILDINGS

FIGURE 3-3: UNCONTROLLED URBAN GROWTH AND DIKE BUIOLDINGS PARTICULARLY CONCENTRATED IN THE ZONE D

Because of this rapid and continued growth, the need to provide proper green infrastructure has occurred at the same rate. However, the provision of this infrastructure, which is necessary to sustain a reasonable quality of life, has not been provided at a rate equal to its demand. This problem has had a tendency to lead to environmental damage, as people have turned to alternative means of compensation for the lack of infrastructure.

Despite the need for improved infrastructure and better growth management, inadequacies in local planning processes have put unprecedented pressure on surrounding locations in zone D, endangering not only long term environmental sustainability but also the quality of life of residents in these areas. In this context, the Etna park is partially similar to other parks in Sicily. Its peculiarity is that it is a volcano. Prior to discussing planning within the Etna regional Park, recent environmental trends in the region are reviewed.

3.1.2 Environmental Planning in the Etna Park

The leapfrog development, which comprises pockets of unconnected development, has led to a fragmented area and a progressive loss of some of the best agricultural lands at the edge of the expanding of various disastrous human interventions, such as flow variations, excavation works and dike building. However, even though this is a well-known and well documented phenomenon, little has been done successfully in terms of planning and policy development in the Etna Park to halt this pattern of development. Additional factors pertaining to past environmental planning initiatives, including politics and government corruption, and the urban morphology and structure of a region such as Sicily must also be taken into consideration.

Foreign and local planners have long been attempting to bring some form of management to the sprawling environmental development. However, this has been only met with limited success. Researchers noted that there had been growing and widespread agreement for over a decade that measures to control the growth of the Etna Park were urgently needed. It was already recognized that the capacity of the land to absorb the needs of the expanding city was diminishing and the correct balance in promoting the safeguarding, running, preservation and defending of the natural environment of the park, like the rest of the parks and preserved areas, was occurring.

The processes of transformation on the zones of the Etna Park have brought changes in settlement patterns, land uses and environmental relations. To study these processes, it is necessary to gain a greater understanding of the underlying spatial and ecological changes.

Integrated within the former is the process of structural change, including transportation systems. In this project, the examination of structural change is regarded as a key element in the determination of land use change and the detection of potential environmental values in order to limit the effect of environmental stress.

By identifying the greenway system change and expansion as one of the primary catalysts of other changes in the zoning,

The concepts described are based on several official documents from the Regional Government Departments, Regional Park Institutions and in particular the Etna Park Institution and may be listed as:

1) The containment and ordering of development particularly within the area of Etna's outer hinterland.

2) The division of development to, and the accelerated growth of, a selected number of satellite towns (such as Nicolosi, Pedara and Mascalucia) and communities.

3) The establishment of a system of major transport corridors in the context of an effective, developed transportation hierarchy in the field of environmental planning.

4) The retention and consolidation of an economically viable, convenient and socially vital "Central Cores" in the field of cultural tourism.

5) The consolidation of existing agricultural locations, typical produce, cultural heritage in general and vegetation.

This list is not exhaustive in terms of the items incorporated into the Park Plan. However, it does represent the main conceptual base of the plan. What is important to note about this is that while representing some obvious development planning principles for any park, the items do not represent the current state of development within the Etna park. The study of any given cross-section throughout the yields evidence to the fact that the planning of Etna Park has not been adequately contained or properly ordered using sound planning practice. The establishment of major green corridors may have been completed to a certain extent. However, these have contributed to exacerbating problems of conservation and valorization in the entire park. Because these plans have been less than successful in many respects, several more recent policy improvement plans have come into effect.

Specifically, more recent policy development has sought to address environmental impacts of urban expansion. In an interim report created by an M.I.T. Consultant Team on strategic planning for the Etna Park, seven objectives on the SEM of the park were proposed. These objectives were:

1) to promote a jobs-housing balance in particular sectors such as agriculture and tourism;

3) to encourage compatible mixed use development;

4) to improve amenities in the park and in particular in zone C and D;

5) to allow for the decentralization of government employment;

6) to assure the installation of infrastructure in advance of development; and

7) to mobilize private expertise and capital in the creation of environmental strategies.

Planning initiatives would require costly green infrastructure and would not aid in rectifying the problem of continuous environmental sprawl in the short term. Second, there were planning initiatives undertaken to rationalize the systems of transit in the park.

Even though many studies have been completed, the list of Etna Park's management outlined by the Etna Regional Park board in terms of achieving the correct balance in valorising and safeguarding of the natural environment of the Park while improving the living conditions and economy of the person within the territory, are still as prominent today as they were almost 16 years ago. It is argued in this research project that the use of spatial information technologies will be able to facilitate improved planning practices such that this list of problems can be more effectively dealt with than in the past. However, M.I.T. Consultant Team notes that there are several shortcomings in the current provincial development planning process that must be overcome. In particular, she lists the following shortcomings:

1) centralization of decision-making and budget control;

2) a lack of cooperation among the government agencies;

3) duplication of data sources and data quality;

4) lack of public participation in planning;

5) the time constraint;

6) plans should respond to the real needs of the local people.

Given these shortcomings, researchers noted that while planning in Catania Province improved systematically since the different territorial programme and plans (PRUSST Plan, PIT Territorial Agreements, P.O.R. Programme, etc.), new planning frameworks and access to data were needed to facilitate a decision making process able to overcome the many existing problems with the Etna Park planning process. Having noted this necessity for improved planning frameworks and data, the future directions of planning practice in the Etna park need to be further explored. The future of planning practice can take many forms. However, several improvements in the facilitation of this practice are necessary before effective planning can take place.

Increases in land use efficiency could be achieved through property and land taxes, which would simultaneously decrease disastrous human interventions and environmental impact fees could help to ensure sustainable and environmental correct growth and ultimately SEM.

Also, Onchan (2002) stated that while a number of policies, plans and initiatives were in place to improve environmental conditions, some of which are discussed in this section, they were still not effective. He attributed their ineffectiveness to inadequate organizational capacity and a complete lack of institutional and governmental mechanisms to respond to the pervasive environmental impacts of the expanding economy and current patterns of social change (Onchan, 2002). Furthermore, researchers noted that in the short term, the Etna Park Executive Committee was to be able to meet the objectives of sustainable development and effective environmental management.

Because of these problems, and because it is important to assume that the Etna Park will implement ISO quality standards (UNI EN ISO 140001 for protected areas) anytime in the near future related to the SD (Puglisi, 2006), it is necessary to determine a means to monitor and assist with the sustainable management of this planning.

As we know, the term sustainability is the basis of the "Green Movement," and understanding the various aspects of the term is the starting point for creating and maintaining sustainable communities. The Piedmont Authority for Regional Planning intends to do that with the Sustainable Communities Regional Planning Project. A key element of the planning process is to engage local citizens and groups who are generally left out of community discussion on sustainability issues. The purpose of this research is to review the methods used to help planners to better understand sustainability within the context of their communities, individual lives and eventually the larger regional community (Powell, R., Bailey, R., Wade, A., and Tunstall, J., 2012).

A means of carefully planned and controlled development is possible and spatial information technologies can assist in achieving this type of planning. The results presented in the following chapter show that the environment can be effectively monitored and analyzed to identify the location of potential environmental stresses inside the Park and in its hinterland.

3.2 Criteria for Selecting Specific Study Area

The study area selected for analysis lies at the Catania Province in the Etna Park. This area is the main gateway to the Etna Park from the airport and is an area of environmental quality and land use change. The site was selected because of the rapid urban development occurring in the area and the high likelihood of identifying stress points and different environmental values.

In particular this study analyses the territory of Catania Province and represents the *test-field* for all possible events related to cultural activities in the Park of Mount Etna volcano or other Italian and European parks for which the geodatabase and thematic maps have been already implemented. The first phase of the job concerns the accurate knowledge of the territory and research related to the definition and the simulation of analysis of environmental management.

The land use in this area is characterized by various cultural activities including, environmental itineraries and cultural buildings (traditional architecture and religious buildings) mixed with idle lands, cultivated lands for agricultural use, and in particular high quality agriculture (IGP and DOP, CEE 2081/92), fauna and preserved vegetation (Lo Tauro 2001 and figure 3-4).

FIGURE 3-4: CULTIVATED LANDS FOR AGRICULTURAL USE, HIGH QUALITY AGRICULTURE (IGP AND DOP, CEE 2081/92), PRESERVED VEGETATION AND CULTURAL BUILDINGS (TRADITIONAL ARCHITECTURE AND RELIGIOUS BUILDINGS)

FIGURE 3-5: IDENTIFICATION OF GREENWAYS, MAIN CULTURAL ITINERARIES AND THE RURAL FRINGE DENSIFICATION OF BUILDINGS AND LAND USE CLASSIFICATION IN CATANIA PROVINCE.

The idle lands are typically left unused for crop or tree planting, as rising land prices partially brought on by volcanic activities (such as the lava flow, volcanic cinders, forest fires).

Part of this territory has led to various disastrous human interventions (such as flow variations, excavation works and dike building) and the diversion of these lands from agriculture to an "unused state" in anticipation of increased value (Weimin, 1995). Because of the activities of real estate speculators, cultural tourism growth, an irreversible trend of agricultural demise in the region is occurring (Greenberg, 1994). Etna Park Executive Committee further noted that due to the concentration of agriculture and tourism activity in the Etna park, there existed a highly skewed distribution of activity here, which is still typical of Catania Province. In general, much of the development noted by Etna Park Executive Committee has evolved in cultural itineraries along the main roads and minor routes and greenways have sprouted from these arterials (Figure 3-5), providing the basic spatial skeleton for an emerging sprawl of multi-use areas.

3.2.1. European environmental policies

The Interreg Programme, financed under the European Regional Development Fund (ERDF), is a Community initiative which aims to stimulate interregional cooperation in the EU. Also the REVER MED project (www.revermed.com), the MACIMED project (Interreg IIIB Medocc, http://www.macimed.org) and the POR provided funding, facilitating management strategies for Greenways Plan. The Sicilian Region, and in particular, the Catania Province and the Etna Park Institution, started to implement some Conservation Assistance Programs designed and facilitated the project's public input process. The 'Green Corridor' area was formally recognized during the Development Plan of the Etna park financed by the P.O.R., when policy makers identified potential zones of cultural growth around and along the green infrastructure (Greenberg, 1994 and Shafer, C.S., 2013) with the support of novel GIS platforms. Numerous individuals from citizen groups, nonprofit organizations, private businesses and local, state and federal governmental units also made significant contributions to the Project. The objective of those programs were to retrace historic trails and roads (such as the *Circumetnea* route) for the heritage conservation and valorization (L.137/02). Although the Public institution's role in the Statewide Greenways GIS Planning Project encompassed a number of different activities, its primary task was to prepare a recommended design for the Informative System. This includes the analysis of all the necessary elements (both cartographic and thematic information) such as thematic map and cartography for ecological sub-system or network and a recreational/cultural network as required by the " La Rete Natura 2000" and "Carta della Natura 2000". In particular the implementation of the Natura 2000 under the The 6th Environmental Action Programme and further implementations encourege projects related to the assessment of the conservation status of habitats and species to halt the loss of biodiversity.

The Law 394/91 establishes legal grounds and criteria for the "Carta della Natura" Programme (www.regione.emiliaromagna.it/paesaggi/europa/carta_n.htm) ralated to environmental values and quality and vulnerability. The project analyses 48 Italian territorial systems under the "CORINE Biotope" guidelines. The project also icludes the detailed definition of the features of the cartographic data at 1: 50.000 scale and 1:250.000 scale and remote sensing sources for the realization of the specialised GIS. The "Natura 2000" programme establishes planning grounds under the supervision of the European

Commission Environment DG's Nature. In addition to the delineation of the Statewide Greenways Planning Project, the Sicilian Region was asked to explore prioritization of ecological networks and to develop a record and maps representing greenway plans, under the P.I.T. Programme, the INTERREG IIC and Forest Programme. The represents the bases for the implementation of European Programme. In particular the National ecologic network is published on the web site http://www.planeco.org and in Naturopa, Council of Europe, 87/1998. Based on the European recommendations by the European Greenways Commission, the Programme incorporates greenways into the responsibilities of the Catania Province and to create a Mediterranean Greenways Coordinating Council (Greenways Council) to assist and advise the Statewide Greenways Planning Project. For this purpose greenways and natural corridor of protected open space were selected in order to manage conservation and recreation activities in Mediterranean countries. The greenways network is characterized by natural land or water features, like ridges or rivers, or human landscape features like abandoned rail and corridors or canals. They link natural reserves and heritage (such as SIC, ZPS), native ecosystems, parks, cultural and historic sites with each other and, in some cases, with populated areas, archaeological sites, UNESCO sites, and cultural heritage in general. We also selected in the project greenways which encompass natural landscape features, such as rivers and floodplains, ridgelines, interior or coastal wetland systems, shorelines, beaches in Catania province. We also considered greenways which encompass a variety of altered or human landscape features, such as utility and abandoned railroad rights-of-way, canals, dikes, and roadside corridors. A proiect GIS was planned to be integrated into the GIS platform of the Greenways System of the Catania Province (fig. 3.4). A GIS model was used to develop an initial analysis of the carthographic sources and techniques for mapping update survey and classifications of greenway network in Catania Province. Then, the project showed many opportunities to suggest changes to that initial plan and in turn, these changes were digitized and the physical plan evolved into a recommended plan. Then, the plan was further modified for the purposes of georeferencing and mapping of cultural heritage, and of improving absolute and relative accuracy. Absolute accuracy refers to the accuracy of features on the map with respect to their actual location on the surface of the earth. Relative accuracy refers to the accuracy of the relationship between features as mapped (Surace, 1998 and ASITA http://www.asita.it). In the start up phase of the project, with a view to the utmost economy and expedition in producing maps for greenways, it was possible however to have poor absolute accuracy and still produce mapped data with a high degree of relative accuracy. To find methods of ensuring the accuracy of both location (the latitude and longitude of a point) and elevation (the altitude above sea level), the proyect requires the GPS technique for mapping update survey and in order to make precise measurements as determined by surveys of a higher accuracy. To draft accuracy specifications from photogrammetry and acquiring information from aerial photographs and satellite image data (remote sensing) the project involves also remote sensing analysis. Sparked by this work, we including studies of map accuracy standards based on the Minerva Programme (http://www.minervaeurope.org/publications.htm) and the The MICHAEL Project. An other important source is the Memorandum of Understanding between Commissioners Wallstrom-Busquin-Solbes of 11 April 2002, on the development of the novel INSPIRE initiatives as a legal initiative of the EU. It addressed technical standards and protocols, organisational, co-ordination and data policy issues, including data access and the creation and maintenance of spatial information. (http://www.ec-gis.org/inspire/).

In this phase we also take into account the more important guidelines (already available *or in itinere*), in order to analyse the modality of the geographic database related to the implementation of the "Testo Unico" (L.137/02) and the documentation produced by the "Intesa Stato-Regioni- Enti Locali" for the realization of the Geographic Information System.

3.3 Field Research

Research, data collection, and preliminary data analysis were implemented with the assistance of the technical staff in the Etna Park Institution and other Regional Institutions, located forty-seven kilometers south of Central Catania. In particular, the Centre for Research on GIS, provided assistance, data and a work area.

3.3.1 RS Data Collection

Several data sources were utilized in order to gather as much RS imagery and ground-based GIS data as possible during the fieldwork period. Imagery available at the Etna Park was used. Other sources were examined at the Catania Province.

The SPOT image of the Etna park from January 18th 1997 was examined first. The SPOT satellite was produced by the CNES (*centre National d'Etudes Spatiales*). The Spot1 was launched February 17th, 1986, the SPOT 2 was launched Jenuary, 1990, and the SPOT 4 was launched in 1997 with the VEGETATION system for the global monitoring of vegetation areas. It supplying 10 metre resolution data on the earth's environment for about ten months. As it orbited the earth in a sun synchronous orbit at an altitude of 832 km, SPOT had a revisit period of 26 days. More than 250 scenes in both multi-spectral and panchromatic sensor modes were acquired over Etna during its operational time. Images of the Etna Park (multi-spectral and panchromatic) are also used in this project. The multi-spectral image contains four channels, two in the visible (0,50-0,59 um and 0,61-0,68 um) and one in the nearinfrared portion of the electromagnetic spectrum (0,79-0,89 um), each with a geometric resolution of 20x20 meters.

To allow for the temporal analysis of land cover change that is essential to the integrated framework presented in Chapter 2, other imagery had to be incorporated with the Landsat and MIVIS images. In this case, a Landsat TM image (February and June, 1995) was used. The Landsat program is the longest running project for the acquisition of imagery of the earth from space. The first Landsat satellite was launched in 1972 and the most recent was launched in April 1999.

Over the past three decades, Landsat satellites have captured millions of images from around the world for research and applications in agriculture, geology, forestry, and regional planning. The Landsat imagery used here has an across-track swath of 185 km

Satellite	Landsat	SPOT
Temporal Resolution (d)	16 - 18	1-26
Equator Eight (km)	705 - 920	832
Radiometric Resolution	8 bit - 7/6 bit	8 bit – 6 bit
Aerial Survey (km2)	185 x 185	60 x 60
Megabyte for Band	40 – 5,3	36 -9

TABLE 3-1: RS IMAGE DATA

at 30-metre ground resolution cell size (except for the thermal band, which has 120 metre resolution).

The Landsat TM sensor collects seven bands of data ranging from a wavelength sequence of 0.45 micrometers in the visible blue through 2.35 micrometers in the mid-infrared portion of the electromagnetic spectrum.

To enhance the temporal analysis, a MIVIS image from 2001 belonging to the INGV is also included in the analysis, caracterised by the 102 different spectral bands of the MIVIS sensor. The MIVIS hyper-spectral scanner is a nodular instrument composed of four spectrometers which simultaneously measure the electromagnetic radiation from the Earth's surface by recording 102 spectral bands:

- 20 in the visible spectral region (0.43-0.83 um) with a range band of 0.02 micron;
- 8 in the near infrared one (1.15-1.55 um) with a range band of 0.05 micron;
- 64 in the middle infrared one (2.0-2.5 um) with a range band of 0.009 micron;
- 10 in the thermal infrared (8.2-12.7 um) with a range band of 0.34 – 0,54 micron;

3.3.2 Global Positioning System (GPS) Image Ground Truthing and Global Navivation Satellite System (GNSS)

To assist in ensuring the positional accuracy of the RS imagery, days of *in-situ* ground truthing were undertaken with the support of the Etna Park staff, the Catania Province staff and private enterprises. GPS (Trimble technologies) belonging to Catania Province ware used for data collection at various positions throughout the test-field. The GPS uses a precision achievable with different parallel channel receptions to track multiple satellites simultaneously and to provide reliable and accurate navigational data. Current positional accuracy with this unit is to within centimetres. More than 20 Ground Control Points (GCPs) - a point on the surface of the earth where both image coordinates (measured in rows and columns) and map coordinates (measured in degrees of latitude and longitude, feet, or meters) can be identified and were collected throughout the study area. Greenway's survey was carried out in collaboration with the research unit of the Catania Province and the Etna Park Staff. The project is continuing the research on these Cultural Itineraries's with positioning technique of moving vehicles (R.T.K., Real Time Kinematic). Differet types of GPS positioning systems were used. In particular the low cost system (Lowrance GPS, Globalnav 200) was used in order the survey some greenway's route.

Those data were integrated in the GIS Platform in collaboration with staff of the "Trimble Sicilia" and Catania Province. An other survey was also carried out with the support of wireless applications.

In particular, novel prototypes utilized smartphone application that allows us to collect and modify data in a disconnected environment. These applications were developed to give users the ability to view and query GIS data, but we had to be within range of a working cellular network tower in order to collect and sync data.

We also tried to collect data while working in a remote location. Then, the data could be easily synced back to the system when we were within range. These new version of wireless applications were equipped with a Sybase SQL database that stored and synchronizes data captured on the smartphone devices. Developing wireless applications with proven mobile middleware enabled tests for disconnected operation, which were particularly helpful to damage assessment crews working in areas struck by voulcanic risks.

Figure 3-6: Geographic Location of Select GCPs Collected with GPS Unit and examples of greenway trajectories surveyed by GPS

We also tried to test most common navigation solutions (either completely or partly) on Global Navigation Satellite Systems (GNSS).

Open source program packages for standard and precise positioning with GNSS. RTKLIB (e.g. RTKLIB http://gplv3.fsf.org/) consist of a portable program library and several application programs (APs) utilizing the library. The features are:

(1) Supports standard and precise positioning algorithms with:

GPS, GLONASS, QZSS and SBAS (Galileo is enabled but not supported in current version)

(2) Supports various positioning modes with GNSS for both real-time and post-processing:

Single-point, DGPS/DGNSS, Kinematic, Static, Moving-baseline, Fixed, PPP-Kinematic *, PPP-Static * and PPP-Fixed

(3) Supports many standard formats and protocols for GNSS:

RINEX 2.10, 2.11, 2.12 OBS/NAV/GNAV/HNAV, RINEX 3.00 OBS/NAV, RINEX 3.00 CLK, RTCM v.2.3, RTCM v.3.1, NTRIP 1.0, RTCA/DO-229C, NMEA 0183, SP3-c, IONEX 1.0, ANTEX 1.3, NGS PCV and EMS 2.0 (see Release Notes for supported RTCM messages)

(4) Supports several GNSS receivers' proprietary messages:

NovAtel: OEM4/V, OEM3, OEMStar, Superstar II, Hemisphere: Eclipse, Crescent, u-blox: LEA-4T, 5T, 6T, JAVAD: GRIL/GREIS, Furuno: GW-10 II/III and SkyTraq: S1315F (see Release Notes for supported messages)

(6) Supports external communication via:

Serial, TCP/IP, NTRIP, local log file (record and playback) and FTP/HTTP (automatic download)

(7) Provides many library functions and APIs for GNSS data processing:

Satellite and navigation system functions, matrix and vector functions, time and string functions, coordinates transformation, input and output functions, debug trace functions, platform dependent functions, positioning models, atmosphere models, antenna models, earth tides models, geoid models, datum transformation, RINEX functions, ephemeris and clock functions, precise ephemeris and clock functions, receiver raw data functions, RTCM functions, solution functions, Google Earth KML converter, SBAS functions, options functions, stream data input and output functions, integer ambiguity resolution,

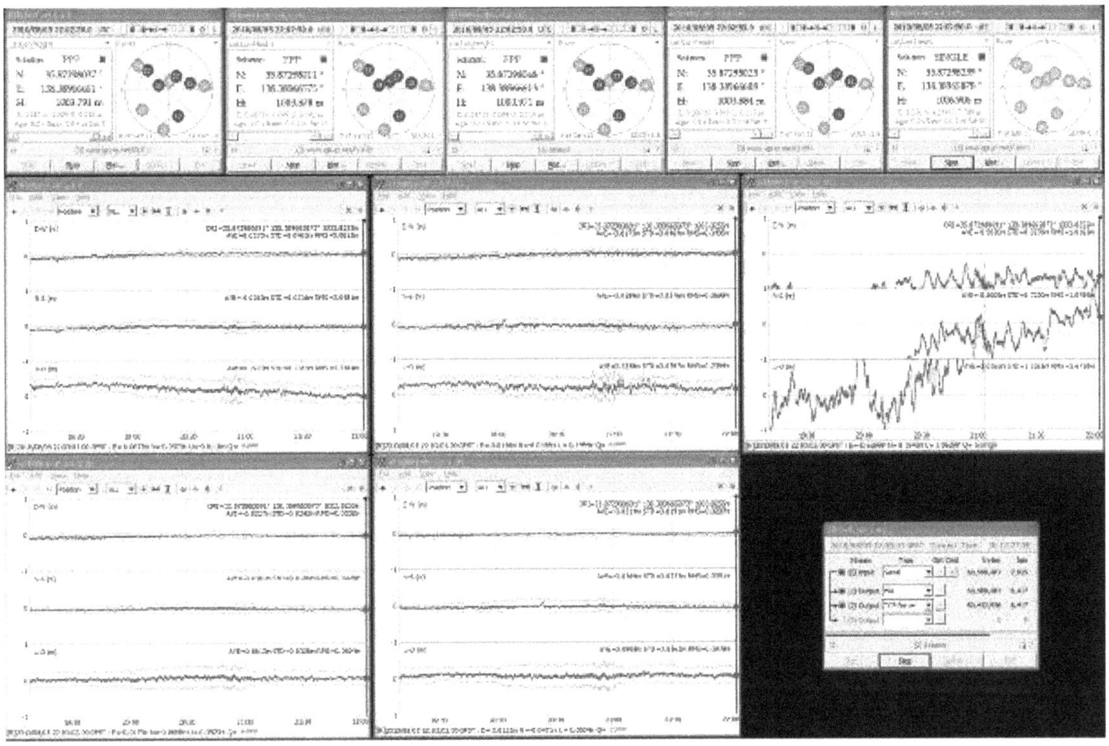

FIGURE 3-6: OPEN SOURCE FOR GNSS POSITIONING

standard positioning, precise positioning, post-processing positioning, stream server functions, RTK server functions

For real-time PPP, EUREF and IGS we have already started to test real-time satellite orbit and clock as RTCM v.3 SSR messages via NTRIP (e.g. registration and account to access the NTRIP broadcasters http://igs.bkg.bund.de/ntrip/index).

However, situations might easily arise where GNSS signal was either not available at all, or did not provide navigation solution with sufficient accuracy e.g., indoors, in narrow greenways. No general solution has been found so far to GNSS-denied navigation, but algorithms which are situation and sensor dependent have been proposed and tested for both personal and vehicle navigation.

Mobil technologies, including smartphones, mobile apps and wireless communications have been tested.

We have been able to collect, view, validate and update spatial data in real time, on location.

In this second phase we get started using the mobile app and the data within GIS Cloud. We installed the mobile app on your Android 1.6+ smartphone device so it worked with GIS Cloud.

3.3.3 Ground-based GIS Data Collection

The collection of ground-based GIS data proved difficult and required a great deal of travel around the Etna Park. Ultimately, data were secured from several sources, including the Etna park Authority, the Regional and National Civil Protection Department, Catania Province Survey Office.

Also, various digital map layers were collected from several sources to facilitate the research. Unfortunately, most of the data collected were poor in quality and no metadata were included to provide information pertaining to data quality, validity, and creation date.

These data layers are summarized in the following items along with their known errors and measures undertaken to render them useful to this research.

- Digital Layer Open Data Source
- Error Source Applied Error Correction
- Administrative Boundaries
    - Incomplete data
    - Minimal Metadata
    - Analysis of coordinate system
    - Geo-registered to National Cartographic Coordinate system
    - Digitized/edited district boundaries where necessary to complete topological analysis of the greenways network.

3.3.4 Field Study Limitations

Several limitations were encountered during the field study period and data (digital and hard-copy) collection process. These include the paper base maps generated by the Catania Province Survey collected for this study were, in most cases, not up-to-date, nor were they in a usable digital format. In general, there is some lack of available digital map resources for the Etna Park at both regional government offices and educational institutions that were adequate in terms of accurateness and completeness. Part of the problem in terms of data availability is the fact that quite often there are many data sources in the park. However, the bureaucracy is such that the different agencies and departments are unable to share their data resources.

Novel tests will give the opportunity to implement innovative solutions. Precise point positioning (PPP) has gained an increasing importance with the availability of precise orbit and clock information, for example from the International GNSS Service (IGS). Whereas PPP was limited in the past to post-processing due to the latency of the precise satellites ephemerides, the increasing number of real-time orbit and clock products opens a broad field of possibilities not only limited to positioning. With the increasing number of global navigation satellite systems, which are currently built-up as alternatives or augmentations to GPS, processing of mixed constellations will gain more importance for PPP (A. Hauschild, 2010).

3.4 RS Data Processing and Analysis

This section outlines the methods of RS data processing undertaken.

Throughout the RS and GIS data processing analysis, several software packages were utilized (ArcGIS 8.2 and ArcView 3.2 – ESRI, 2000 and ENVI 3.5 - RSI) in order to explore and identify the specific software and the specific analytic components used. Where possible, software neutral terminology is used throughout the remainder of the project. An important process related to the integration of RS and GIS technologies concerns spatial data creation. In most cases, the mapping phase, which can be achieved either through automated feature extraction processes or via manual digitisation, of converting satellite image information into GIS data is on of the most time consuming and expensive component of building a multi-source GIS database. To achieve this, several analytic steps were undertaken with the temporal images to study land use change over time as well as to explore greenway feature identification and extraction.

These analysis steps included:

1. Geometric Corrections

2. Change Detection

o Training area classification

o Results Exportation

3. Linear Feature Extraction

o Results Exportation

The geometric correction and change detection steps are now discussed in detail. Following the novel GIS analysis (Abe A. et al 2012), the RS linear feature extraction is discussed.

3.4.1 Geometric Corrections

Merging multi-sensor image data is a commonly used procedure in the study of land use and land cover change detection (Shaban and Dikshit, 2002, Tziztiki, J. G. M., et al. 2012).

Thus, the first step in the analysis is the geometric correction of the Landsat and SPOT Data and MIVIS images.

Existing, albeit inadequate, street and administrative boundary digital GIS layers were used to provide partial ground control points for the geo-referencing process. Also, coordinates collected with the GPS receiver were used to provide additional positional accuracy. The images were corrected to the Universal Transverse Mercator (UTM) grid coordinate system, with the World Geodetic System 1984, (WGS 84) Ellipsoid projection. It was important that the images were geo-registered to the coverages in order to achieve successful future integration and analyses with ancillary ground-based GIS data. The images were georeferenced to the Etna park road and administrative boundary digital map layers, which were in ArcGIS coverage format. Once the first image was geo-referenced, all subsequent images were geo-referenced to this.

The geo-referencing process involved a measure of GCP distortion as it computes the square root of the squared deviations to represent a measure of which GCP exhibit the greatest error while the model order determines the order of polynomial used to perform the registration warping. In general, the higher the order of polynomial transformation, the greater the number of GCPs required for registration and the more complex the warping that can be achieved. Following registration, all of the images were clipped to the boundary of the combined area of Catania Province.

3.4.2 Change Detection

Several approaches were investigated for image analysis. First, change detection analysis was completed for both the Landsat and the SPOT PAN RS images. The MIVIS image from 2001 was not used in this analysis as the Landsat image acquired in 1995 (February and June) and the SPOT PAN acquired in 1996 (August) provided the maximum temporal range for land use change detection. These approaches, also including image overlay, training areas, NDVI analysis, methods of classification and Region of Interest (ROI) creation and Density Slides and the orthorectification model, provided a good account of data. Classification and variability between different surface reflectances and thus successfully distinguished areas of change were defined, particularly at the zone C and D. Of these methods, the training area method was determined to be one of the most effective in terms of visually highlighting areas of change and was therefore carried forward in the analysis. In raw image differencing, co-registered images from different dates should be subtracted pixel by pixel. The resulting digital image represents then change between the two images with different bands as the input images. This process results in positive and negative values in areas of radiance change and values of zero in the areas of no change in a new image. In this project, the Landsat image was compared with the SPOT image for land use detection and finally with the MIVIS image. A portion of the result of this analysis is shown in figure 3-8. The best combination of bands, based on visual interpretation, was selected for further inclusion in the analysis.

White = No Change; Gray = Change / No Data (outside denoted boundaries)

Interpretation of images can be generally achieved by visual inspection as grey tones represent areas of little or no change and areas of very bright or very dark, white and black respectively, as areas of temporal differences. However, some of these differences in the landscape areas of the study site can be attributed to differences in crops, time of year (seasonal variation) and crop rotation practices.

A semi-automatic classification was also performed on the MIVIS image. This method was used to identify homogeneous areas of change thereby separating a region of change from a region of unchanged land cover. This supervised classification produced a thematic map (change/no change) image of the study area based on the variation of vegetation as a results of the change detection process.

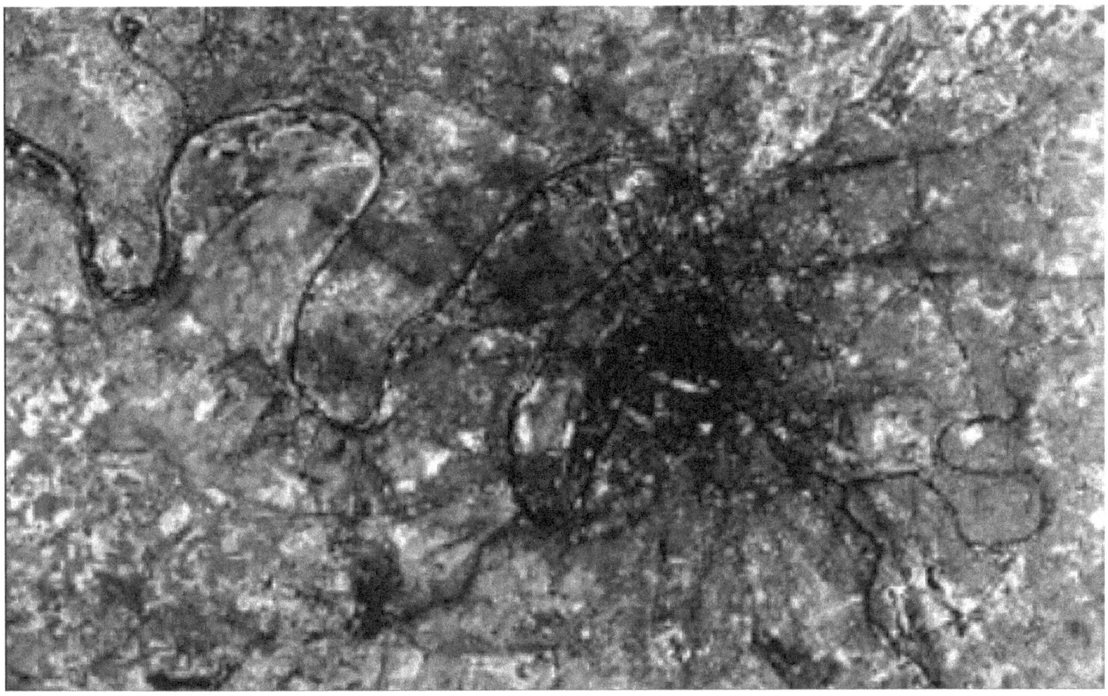

FIGURE 3-7: CHANGE DETECTION IMAGE, REAL COLUR AND UREAL COLOUR (LANDSAT)

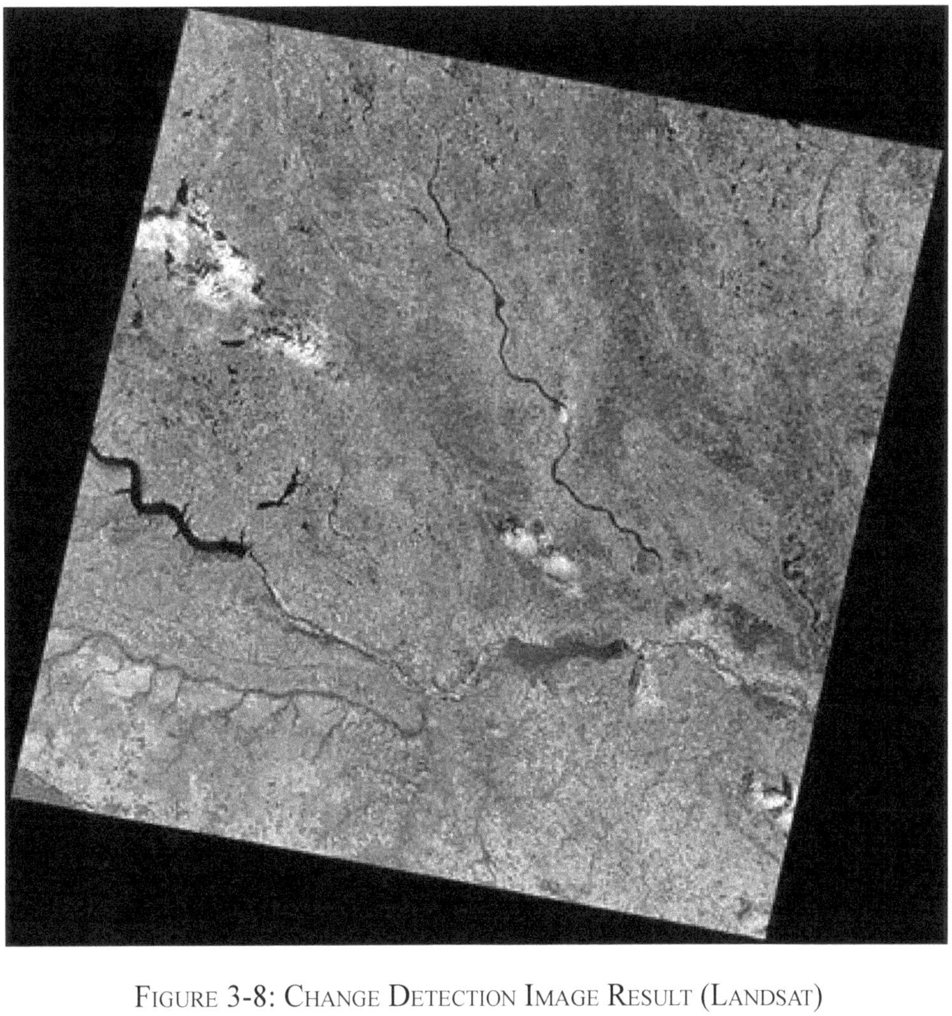

FIGURE 3-8: CHANGE DETECTION IMAGE RESULT (LANDSAT)

3.5 GIS Data Processing and Analysis

In conjunction with the satellite image analysis, other methods were mainly used to identify environmental stresses on the zone C and D of the study area. The initial steps undertaken in the GIS data processing included:

1) Geometric Correction of Digital Layers

2) Data Input

3) Data Conversion

4) Data Relates

The following sections outline the steps undertaken to input, correct, and analyze the tabular data.

3.5.1 Geometric Correction of Digital Layers

The first step in the GIS process was to geo-register the digital data layers to the corrected digital layer. This was achieved by defining the data projection and subsequently transforming the administrative boundary layer as necessary to facilitate future integration and overlay analysis. The UTM WGS 84 coordinate system, the same as that applied to the RS images, was applied to the GIS data layers.

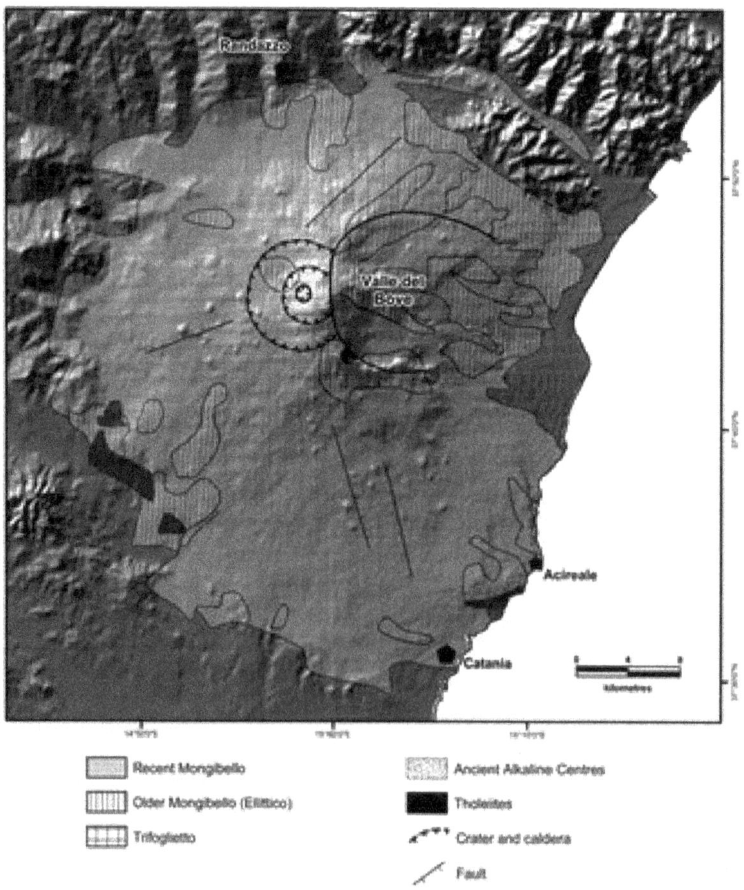

FIG. 3.9: DEM (10 METERS) WAS DERIVED BY THE OFFICIAL CARTOGRAPHY OF CATANIA PROVINCE (1999) IN DWG, SC. 1:10 000.

In particular The LANDSAT and SPOT image were also geo-referenced ed to the UTM WGS 84 system and, in a second phase, orthorectified in the National Reference System (Gauss-Boaga) in order to allow the correct overlay analysis with the suitable cartography. During the orthoprojecting phases a DEM (10 meters) was derived by the official cartography of Catania Province (1999) in DWG, sc. 1:10 000.

An other important source was the DEM of the IT2000 programme of the "Compagnia Italiana Ripreseaeree" (40 m, with about 5 m. of accuracy).

3.5.2 Data Input and Conversion

This section outlines the analysis of areas of potential environmental stress. The incorporation of data is an important step in locating and analysing trigger points at the different zones of the Etna Park. Different parameters were considered in this research: net of tourisms routes, acreage of land use /land cover under urban and rural use, number of rural dwellings and number of registered building under various disastrous human interventions, the greenway network. All of these indictors were discussed in Chapter 2 (figure 2-4 and figure 2-5).

Land use data were entered from printed tables and a locational relation was created from the Provincial and District codes to facilitate their integration in subsequent stages of the analysis process. The finished spreadsheets of data were converted to shapefile format to allow import of the data into the GIS software.

3.6 RS, GIS Integration and Analysis

This section outlines the process of integration of the satellite imagery with the ancillary ground-based data within the GIS to detect and localize areas of potential environmental values and stress along the different zones in the Etna Park.

These analysis steps included:

1. Image to ArcGIS - GRID Conversion

2. Buffer and RS Analysis

3. RS Change Detection

4. The semi-authomatic approach

5. Localization of Stress Areas and Definition of Environmental Policy

Each of these stages of analysis is outlined in further detail below.

3.6.1 Image to GRID Conversion and Vectorization

The first step in the conversion from RS processes to the multi-source GIS database was to export the RS image files into GIS-ready format. The derived areas of land use change (4) were exported as shapefiles. This format includes:

1) supported by GIS software

2) may contain geo-referencing information in header tags

3) supports black-and-white, grayscale, pseudocolour and true colour images.

The shapefile containing the classification of the change detection results was then converted to ArcGis - GRID format. The image was converted to a GRID as an interim process as raster images can be more easily integrated in a multi-source GIS database. Subsequent to GRID conversion, the original image file was integrated and converted in ArcGIS to a vector polygon coverage based on the change and no change classes. This was achieved with the ArcGis 'gridpoly' function in order to facilitate the integration of the land use change data derived from the RS imagery with the ground-based greenway network data through a tabular join.

Following the creation of the New Maps, we added layers to the project from several sources: local computer, Open source database, Web Map Service or a Tile Map Service such as Open Street Maps.

New layers from a local computer habe been uploaded using the File Manager. In File Manager we selected a directory to which we uploaded and selected Upload. Then, we selected browse to upload files.

3.6.2 Greenway Buffer and RS Training Area Detection

To link further the results of the land use RS change detection analysis and the brounfild analysis with the major green infrastructure, a process of proximity analysis was undertaken at the provincial level. Using the existing major road network digital layer major roads for cultural itineraries and greenways in the provinces of Catania were buffered to distances of 100, 200 and 500 metres (Figure 3-10).

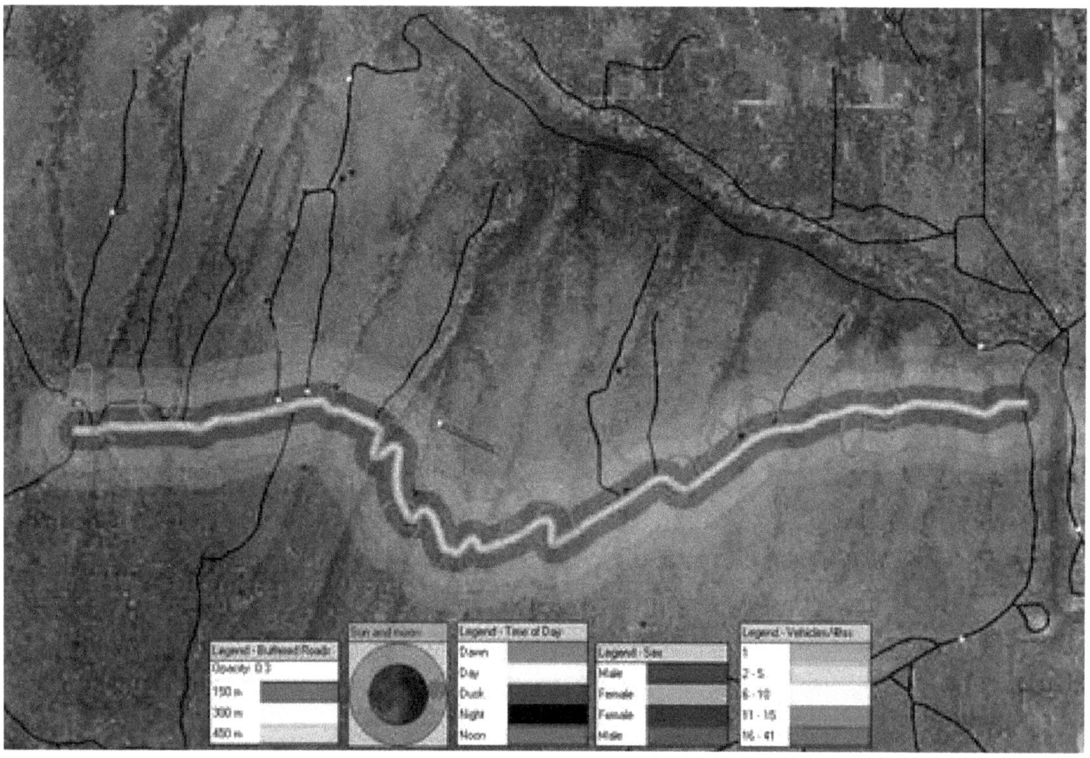

FIGURE 3-10: MAJOR GREENWAYS WITH 100 METRE BUFFER

These buffered areas were then overlain with the semi-authomatic classification of the RS detection analysis to determine the proximity of detected change to the existing green network using the MIVIS Images.

Analysing the radiometric content of the pixels belonging to different zones of the satellite images (greenways, streets, vegetation, etc.) was defined a different reaction of the greenway paving system (land, bitumen, etc) with different values of radiance.

In a second phase buffered areas were also overlain with the *training areas* in order to determine the proximity of detected change to the existing green network. The definition of the training areas was particular important because it defined a classification system for small areas related to the greenway network and wide and heterogeneous areas such as sown field, land, ground, vegetation etc. The training areas had to represent clusters of values related to the analysed class. I defined twenty *training areas* related to the land use. Five training areas were also defined for the greenway network.

This phases also deals with the process of classifying multispectral images into patterns of varying gray or assigned colors that represent either clusters of statistically different sets of multiband data (radiances expressed by their DN values), some of which can be correlated with separable classes/features/materials (Unsupervised Classification), or numerical discriminators composed of these sets of data that have been grouped and specified by associating each with a particular class, whose identity is known independently and which has representative areas (training areas) within the image where that class is located (Supervised Classification).

In the *unsupervised classification* individual pixel is compared to each discrete cluster to analyse which one it is closest to. A map of all pixels in the image, classified as to which cluster each pixel is most likely to belong, is produced (in black and white and in colors assigned to each cluster). This then was interpreted as to what the color patterns may mean in terms of classes, etc. that are actually present in the real world scene (scene's feature/class/material content). In my these I particularly focussed my attention to the supervised classification. In the *supervised classification* I knew beforehand what classes, etc. were present and where each was in one or more locations within the scene. These were located on the image, areas containing examples of the class were circumscribed (making them training sites), and the statistical analysis was performed on the multiband data for each such class. Instead of clusters then, one has class groupings with appropriate discriminant functions that distinguish each (more than one class had similar spectral values using more than 3 bands because different classes/materials seldom had similar responses over a wide range of wavelengths).

All pixels in the image lying outside training sites were then compared with the class discriminants, with each being assigned to the class it was closest to - this made a map of established classes (with a few pixels remaining unknown) which could be reasonably accurate (but some classes present were not set up; and some pixels were misclassified).

It is pointed out that many of the areas classified in the image by their cluster values may or may not relate to real classes (misclassification is a common problem).

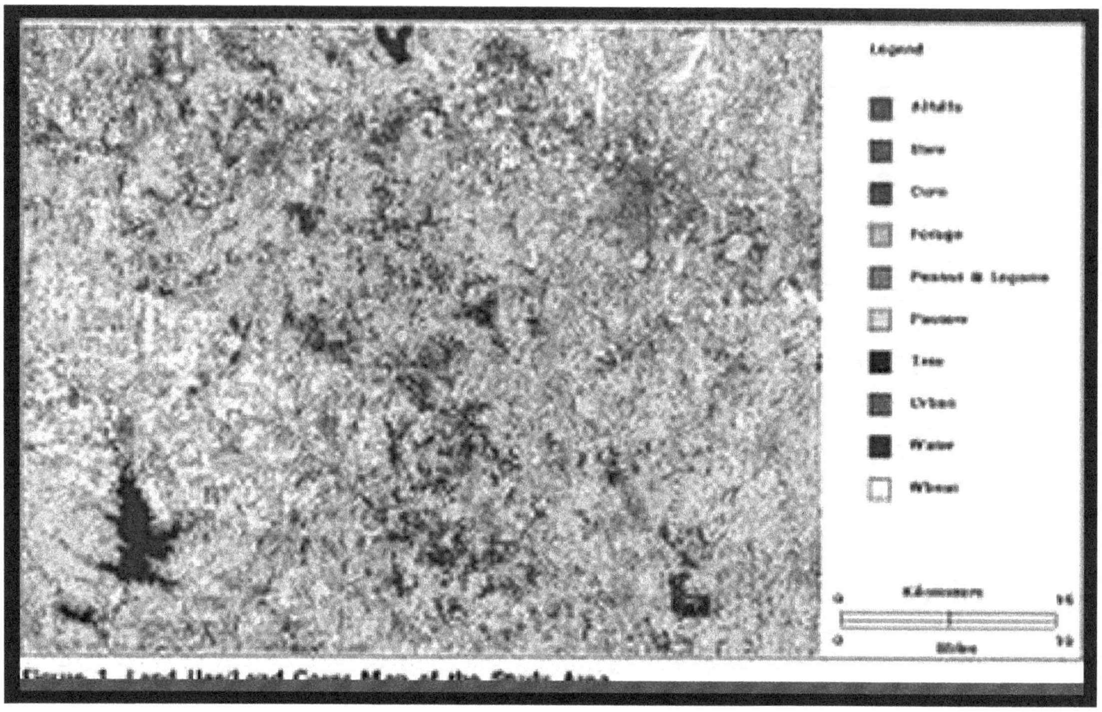

FIG. 3.11: THE SUPERVISED CLASSIFICATION, "MAXIMUM LIKELIHOOD" AND ROI IN CATANIA PROVINCE

3.7 Localization of Stress Areas & Growth Management Policy

The methods identified above were used to detect potential areas of environmental stress in the study area.

In particular, the remote sensing detection techniques for data integration in GIS aimed at monitoring environmental stress and brownfields normally generated by abandoned areas, degradation of the fitosanitary status of the vegetation and buildings illegally built. In reticular, hazard increases when dangerous materials are abandoned in brownfields. This analysis detects environmental data by means of the airborne MIVIS sensor in order to allow me a detailed classification of the territory and the identification of potential areas of environmental stress or brownfields.

In fact a valuable contribution to environmental stress surveys was made by high spectral resolution spatial data processing. Particularly, research work dealed with use of hyperspectral sensors, capable of extending the remit beyond the visible zone of the electro-magnetic spectrum till the thermical infrared area too. The presence of structures, building demolition and dangerous materials in the top subsurface is likely to cause variations in humidity in the surface. These variations affect both vegetation, and some physical features of the ground such as thermal conductivity and capacity. Especially in the first hours of day, you can notice thermal anomalies due to different evaporation. The exam of these anomalies, carried out by the use of techniques of digital processing of images in the spectrum bands particularly sensitive to the above mentioned indicators, enabled to determine possible signs of environmental stress and brownfiels. An experimentation of these hyperspectral techniques was conducted on Catania province and the Etna park using digital images (already available) produced from MIVIS (Multispectral Infrared and Visible Imaging Spectrometer) airborne system, taking simultaneously the radiation coming from the ground surface covering totally 102 bands of electro-magnetical spectrum (from visible to thermal infrared area). In this particular case I analysed the 92-93 and

94 bands in the thermal infrared area. The final product of this experimentation was a thematic map expressing the possibility to determine the areas with the highest potential for environmental stress.

Once the localization of potential stress areas was complete, these areas were reviewed for use in the temporal linear feature extraction analysis based on their ranking. Ultimately, one area was selected for continued analysis due to its high overall ranking indicating a strong potential for environmental stress. Based on the results, attempts could be made to define new growth management policy. Policy development, in a wider sense, includes both strategic policy and operational budget development. At the centre of the emergent growth management policy is the fact that identified areas of potential stress allow planning practice to work toward achieving SEM.

Specifically, problems associated with duplication of data sources and, in particular, data quality can be improved through the integration of RS, GIS, and "open source data" to facilitate environmental stress indicator analysis. Moreover, GIS tools correlate and fuse many data in such a way to evaluate extension and derivative parameters of the same brownfields and to locate indicators that can be used within the framework of a sustainable planning and development (Report for the GMES Initial Period, http://www.gmes.info/)

With timely and accurate information about the changing morphology in Catania Province, the difficult process of planning for current and future growth and subsequent land use change will hopefully be mitigated.

3.8 Linear Feature Extraction

Edge detection is the principal means of detecting from imagery greenways and other linear features that have definite boundary edges, as discussed in Chapter 2. If homogeneity can be found within a certain region bounded by a sharp, contrasting periphery, this represents the edge of this particular feature. In this case, specific edge detection algorithms included in the Image Processing Toolbox of the ENVI software package were utilized to attempt to isolate the greenway network from the satellite imagery. In this portion of the analysis, only the area termed as the Green Corridors were analysed. This limited geographical focus was used because if green infrastructure could be isolated within this area, then they could also be isolated in the broader study area. In the project it was deemed the best location to test the accuracy and effectiveness of the linear feature extraction process outlined below.

To enhance the imagery further prior to applying the edge detection algorithms, supervised algorithms were applied. In particular, the definition of the *training areas* offered the opportunity to classify satellite images on the base of the supervised algorithms. After different tests I realise that the best result are related to the implementation of the *maximum likelihood* algorithm.

Several edge detection algorithms are available for researchers such as the Sobel, the Laplacian of Gaussian, and the Canny algorithms. The Sobel algorithm performs a spatial gradient measurement of an image using a pair of 3 x 3 convolution kernels to emphasize regions of high spatial frequency, which are equivalent to edges (Fisher *et al.* 2000, Sharma, et al., 2013)

Convolution can be defined as means of multiplying two arrays of numbers of different sizes to produce a third array of numbers. For example, a small matrix, also known as a kernel, could be multiplied together to produce a third matrix of numbers in a convolution. Similarly, the Laplacian algorithm is used to highlight regions of rapid intensity change using a convolution filter (Fisher *et al.* 2000, Sharma, et al., 2013)

The Canny edge detector algorithm is also designed to be an optimal edge detector that works on a multi-stage process. First, the image is smoothed using a Gaussian convolution. This step is used to blur the image to remove unnecessary noise.

Subsequently, an operation is applied to the image to highlight regions of the image with high contrast. These contrasts are representative of edges, as they often correspond to high magnitude gradients in the image.

The ENVI software provides three supervised algorithms of classification: *maximum likelihood*, the classificatory with maximum verisimilitude; *minimum distance; parallelepiped,* the classificatory with box type. I should say that only the *maximum likelihood* method was applied to all the satellite images. It adopts decisional criterium on probabilistic typologies. That algorithm calculates the probability to belong to the different classes for each pixel and than it apply to those classes with high probability.

After different tests, I utilised different statistic values in this analysis. In particular I selected the 80% values for the bitumen typologies while the 97% for other classes such as sown field, land, ground, vegetation etc. This choice can be related to the individuation of the greenway network. In particular I preferred to consider pixels that belong to other classes rather than to neglect pixels that belong to the bitumen typologies.

Considering the heterogeneity of the different classes I preferred to utilise all the channels available at the same time. The results of each analysis was visually inspected for the clarity of the extracted greenway features from the different types of imagery under examination. The results were evaluated and accordingly accepted or rejected. If accepted, the extracted greenway segments were kept and advanced to the next stage in the overall methodology. If rejected, alternative algorithms should be iteratively tested until the best extraction results will be achieved.

3.8.1 Analysis, method and thematic maps

Thematic maps (such as mine and quarry positions, litotype characterization, vegetation cover and types) were analysed exploiting the 102 different spectral bands of the MIVIS sensor.

In this phase I also applied the NDVI and textural analysis to the same scene, in particular using this time the NIR (near-infrared) band in order to define disomegeneity in the component of the vegetation (Donnay et al., 2001 and Borowik, T., 2013).

The aim is that of identify of some possible vegetation differences (in term of its components) in the same areas before identify as uniform. The methodology establishes a prior semi-automatic approach for the research on vegetation which is suggested to become an ecological corridor (usually a narrow structure of vegetation) in order to connect areas with peculiar naturalistic and environmental quality.

Particular attention has been paid to the calibration of the sensor, extraction of the spectral signatures for the classification of the greenway network, and to the geo-regerencing of the images. Such operation was done orthoprojecting the images through the sensor procedure, implemented within the ENVI programming language and softwares.

3.8.2 Significant band selection and the semi-automatic approach

An important step of the project was to select the most useful bands for the detection of greenways features and vegetation anomalies. These can be identified by the NDVI and by the texture, soil mosture and vegetation cover differences that are produced by greenway network.

In my case-study, the peculiarity of the investigate objects led me to select bands renouncing to the principal components analysis; I peceeded with visual analysis interpretation taking care of the bibliographic references. A total of about 10 of the 102 available bands of the MIVIS sensor were chosen.

4 in the visible range (b2=0.4600 um, B7=0.5600 um, b211=0.6400 um, b20= 0.8200 um) for the contextual location;

2 in the near infrared range (b23= 1.2750 um, b 28=15250 um) for vegetation cover anomalies;

1 in the medium infrared range (b 52 = 2.1790 um) for soil moisture;

3 in the thermal infrared range (b93= 8.3859 um, b97= 10.0200 um, b101= 11.9450 um) for thermal variation on the ground.

An other step of this project involved NDVI (*Normalized Difference Vegetation Index*) analysis.

The NDVI is a model for converting satellite-based measurements into surface vegetation types. The NDVI uses a complex ratio of reflectance in the red and near-infrared portions of the spectrum to accomplish this. Reflectance in the red region decreases with increasing chlorophyll content of the plant canopy, while reflectance in the infrared increases with increasing wet plant biomass.

The 2001 MIVIS Image has been classified based on the NDVI by measuring the wavelengths and intensity of visible and near-infrared light reflected by the land surface.

Very low values of NDVI (0.1 and below) correspond to barren areas of rock, sand, or snow. Moderate values represent shrub and grassland (0.2 to 0.3), while high values indicate temperate and rainforests (0.6 to 0.8).

It was used an algorithm called a “Vegetation Index” to quantify the concentrations of green leaf vegetation around the Etna Park. It was also possible to create detailed maps of the green vegetation density that identify where plants are thriving and where they are under stress (i.e., due to lack of water). NDVI is calculated from the visible and near-infrared light reflected by vegetation. Healthy vegetation (left) absorbs most of the visible light that hits it, and reflects a large portion of the near-infrared light. Unhealthy or sparse vegetation (right) reflects more visible light and less near-infrared light. The numbers on the figure above are representative of actual values, but real vegetation is much more varied.

Nearly all satellite Vegetation Indices employ this difference formula to quantify the density of plant growth — near-infrared radiation minus visible radiation divided by near-infrared radiation plus visible radiation (Gomarasca, 1997). The result of this formula is called the Normalized Difference Vegetation Index (NDVI). Written mathematically, the formula is:

$NDVI = (P_{IR} — P_R)/(P_{IR} + P_R)$

Were :

P_{IR} = reflectivity in the near-infrared radiation (0,76 – 0,90 um)

P_R = reflectivity in the red radiation (0,63 – 0,69 um)

Calculations of NDVI for a given pixel resulted in a number that ranges from minus one (-1) to plus one (+1); however, no green leaves showed a value close to zero. In fact, a zero means no vegetation and close to +1 (0.8 - 0.9) indicates the highest possible density of green leaves (Colwell 1974, Gomarasca 1997, and Borowik, T., 2013).

In a second phase the NDVI values are also classified in three intervals using the "quantile" algorithm which gives the same number of observations to each class, such as the following ones:

class 1 – absence of vegetation, NDVI <= 0.15;

class 2 – vegetation with medium values, 0.15 < NDVI <= 0.5;

class 3 –vegetation with high values, NDVI > 0.5,

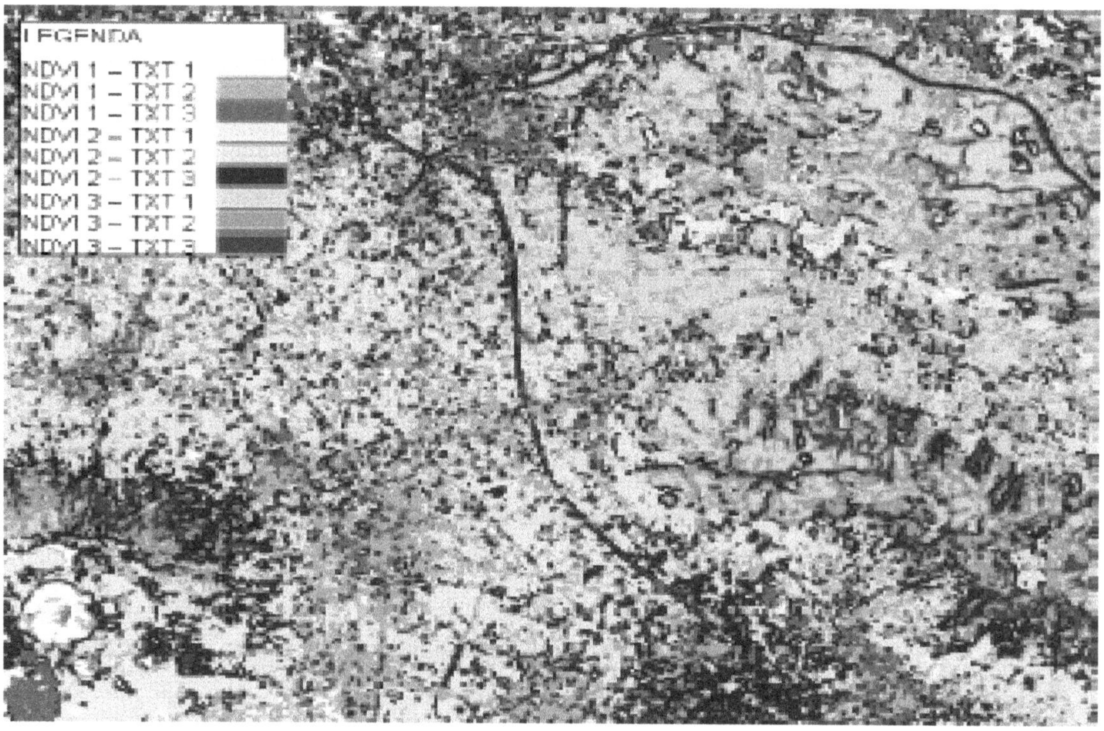

FIG. 3.12: TEMATIC MAP AS THE RESULTS OF NDVI ANALYSIS AND NIR BAND

Additive synthesis in real colour RGB is also elaborated from the combination of the band 13 (0.673-0,693 micron- productivity of the soil), the band 7 (0,553 – 0,573, 2 micron – organic component in the soil), and band 2 (0,453-0,473 um – yellow component in the water), and the near-infrared 19 band (0,793-0,813 um – structure of the vegetation), band 13 (0.673-0,693 um), band 7 (0,553 – 0,573, 2 um) of the MIVIS sensor.

The elaboration of the false colour image gives the opportunity to map the environmental and man-made structures with different tones, and in particular the following ones: the red tones for the vegetation; the blue tones for the building structures, the land with absence of vegetation in brown. This typologies of tonalities are used in the mapping of vegetation because generally shows the condition of hydration of the land of the environment and the vegetation anomalies. Generally the dark red tones define vegetation with good health (or vegetation with major quantity of chlorophyl) while orange and yellow tones show the presence of minor vegetal biomass (or vegetation with minor quantity of chlorophyll.

Such kind of analysis gives information on the frequency and disposition of the variation in the tones in particular zones of the image. This analysis gives also a first estimation of the minor/ major regularity in the vegetation related to the localization of urban areas in the territory.

Different regions of interest (ROI) were selected inside the sample areas: buildings, streets, soil moisture, orchards, vegetated fields, non vegetated fields, water, shadow zones.

Rule images that were generated during the classification were considered. In rule images the dark pixels means a similar spectral signature to the selected class, while the gray scale pixels mean different one.

The overlay of the defined thematic map with the "Map of Vegetation" of the Mount Etna of the Institute of Biology and Vegetal Ecology, University of Catania, sc 1:50 000. This phases of analysis gave the opportunity to define classes (absence of vegetation, vegetation with medium values, vegetation with high values of NDVI) that were non presented in the Map of Vegetation. This phases also applied the detailed analysis of those sources with the detailed land cover/vegeation map (one to four level of detailed CORINE, see chapter two).

In fact it is well known that the updating of a reliable cartography, both for its metric quality and its contents, represents an indispensable tool for the realization of a specialised GIS. Furthermore, this necessity becomes more important in the field of conservation and valorization of cultural sources. In fact natural factors (and in particular the volcanic risks) can deeply modify the morphology of the land use and the characteristics of urban areas and natural environments.

The primary objective of this phase of analysis will be the updating of the cartographic data, and of the informative contents (presented in the database) which are strictly connected to each other. The main purpose is to avoid useless duplications and redundancies (on one side and informative errors and incompleteness, on the other) and to stimulate the implementation of multi-scale and multi purposes systems.

For this aim, the MIVIS image resulted geo-referenced and orthorectified to the National Reference System in order to allow the correct overlay analysis with the suitable cartography using the CTR. Rule images were very helpful in highlighting the presence

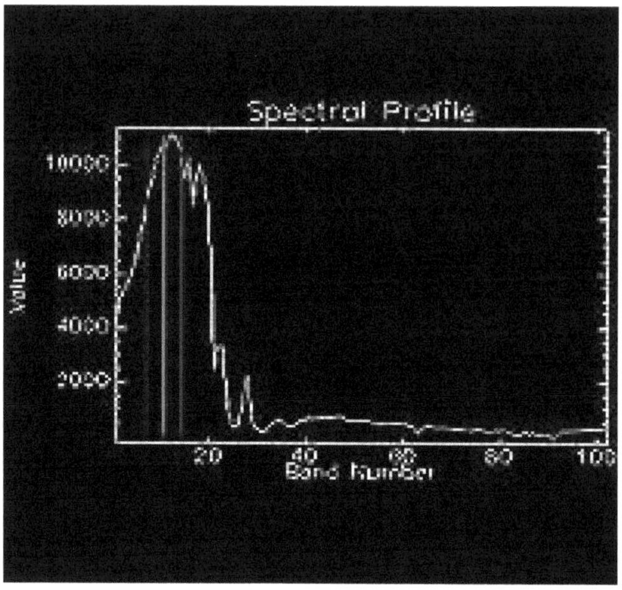

FIG. 3.13: SPECTRAL FEATURES ANALYSED BY THE DOWER-SPECTROSCOPY-LAB.

of typologies of greenways on the bases o different spectral characteristics of the ground. The rule image of different classes, such as "bitumen", "earthware" etc, enabled us to recognize linear anomalies belonging to the greenway network. It was also possible to classify the different typologies of greenways comparing those different spectral characteristics of the ground with the spectral characteristics analysed by the Dower-Spectroscopy-Lab (http://speclab.cr.usgs.gov) and other "open data-base".

For the classification of the green infrastructure I utilised the standard for the monitoring and realisation of greenways of regional interest (Law 86/1983).

Other sources such as aerial photography and ortophotos has revealed the wealth of hitorical greenways that is one of the fundamental resources for my undertsnding the connection between cultural landscape through time (diachronic variations). In particular the methodology of aerial photography is well established and the underlying reasons for the appearence of historical sites through cropmark and soilmark evidence is well understood (Wilson 1982, 2000; Scollar et alii 1990, see Musson C., Palmer R., and Campana S, 2010). Some tests has been carryed out during the XIV Summer School "Digital Arial Photography and GIS based interpretation" , Siena and Grosseto University, in collaboration with LAP&T, British Cultural Heritage and AARG (Arial Archeology Research Group) under the Cultura Programme.

The devises passively recorded reflected visible and non-visible solar radiation in specific spectral bands (such as the NIR band) able to show vegetation stress, soil moisture variation etc. through which greenways were identified.

An other update for the Automatic Classification is the selection of "open data" for QGIS

The major implementation from the previous version are:

- Improved interface layout;
- The plugin can calculate the spectral signatures of ROIs;

- A new “Spectral signature” tab allows for the visualization of the spectral signatures in a chart;
- A new “Band set” tab allows for the definition of a group of single bands as image input.

In particular, the interface is now composed of 5 tabs:

- “ROI tool”, for the creation of ROIs;
- “Classification”, for the preview and classification of the image;
- “Spectral signature”, for the visualization of ROI spectral signatures in a chart;
- “Band set”, for the definition of a group of single bands as classification input;
- “Settings”, for the advanced settings.

Also, a progress bar is shown during the main processes; a “Cancel” button allows the user to stop the process, but the process is stopped after the completion of the very last Python command.

It is also possible to calculate the spectral signature during the ROI definition.

ROI statistics (Mean and Standard Deviation for each band) are stored in the shapefile, which is created with the required fields according to the number of raster bands. Also, the plugin can add the required signature fields to any existing shapefile.

The Automatic Classification Plugin can work with multispectral images, also if the bands thereof are separated files (single band images). The “Band set” tab allows for the definition of a group of images (i.e. a band set) loaded in QGIS, which is used as input image for the ROI creation and the classification processes.

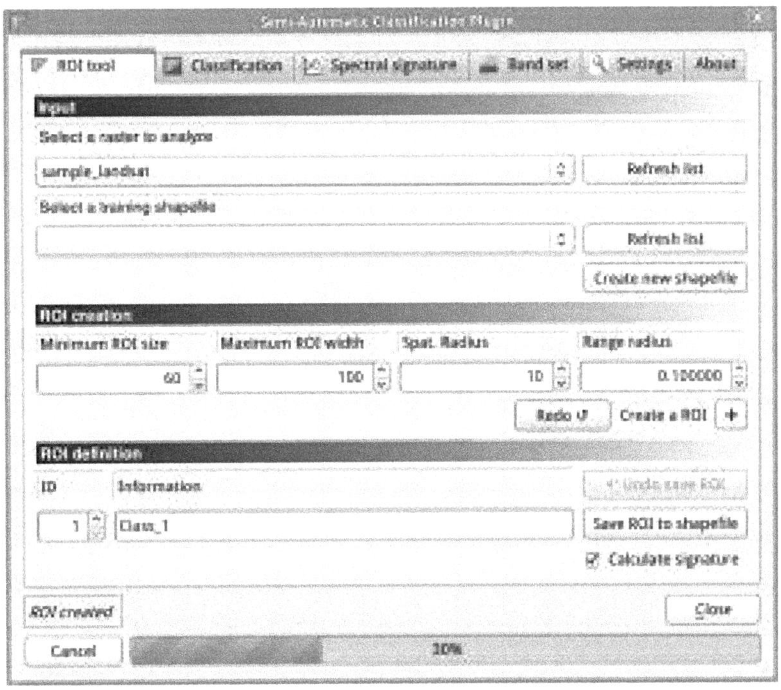

SEMI-AUTOMATIC CLASSIFICATION PLUGIN INTERFACE

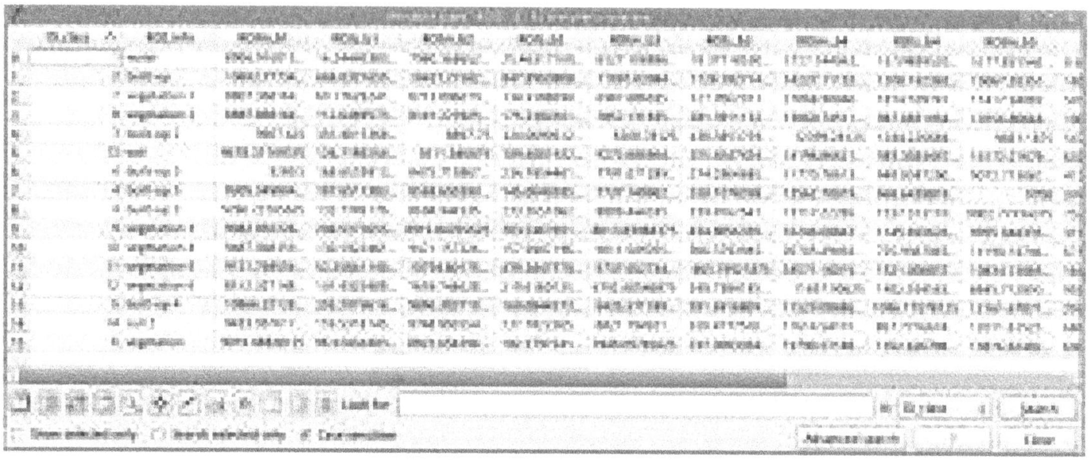

ATTRIBUTE TABLE OF A SHAPEFILE WITH ROI STATISTICS

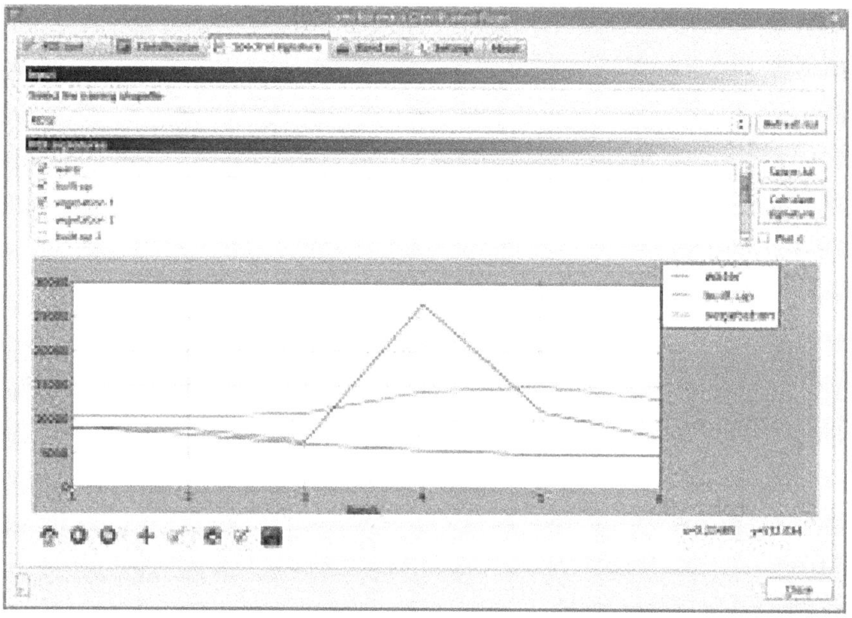

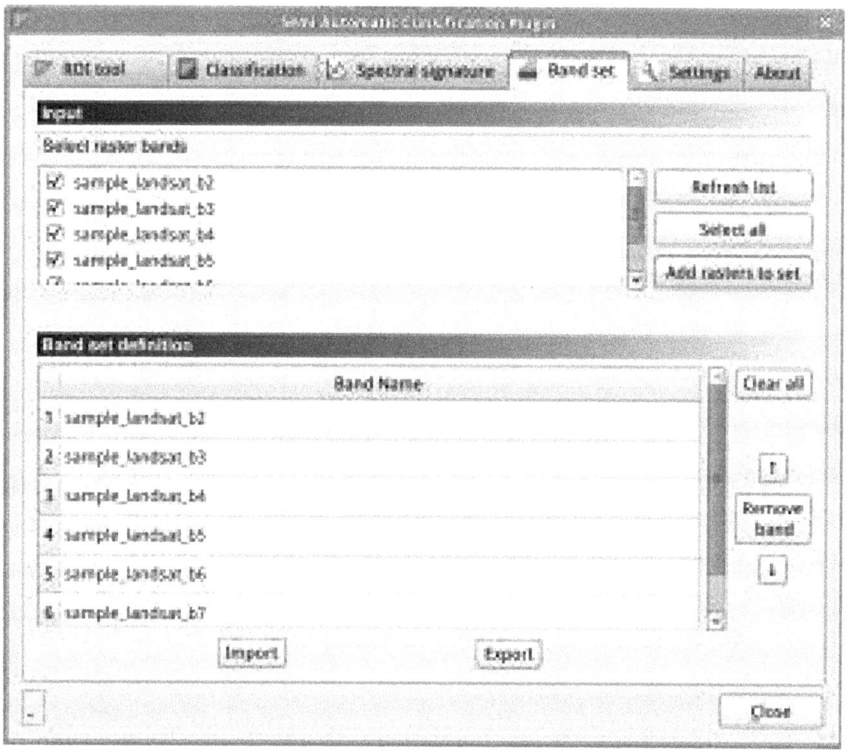

This is especially useful for hyperspectral images, because the classification process is rapider with separated bands. Band sets can be saved and imported, for the purpose of testing the classification results with different band compositions.

If a band set is loaded, the item << band set >> appears in the raster lists of the "ROI tool" and "Classification" tabs.

Those information gives also the opportunity to offer a method of classification of new green infrastructure in order to link areas with peculiar characteristic (such as SIC and ZPS and cultural landscape in general). This study attempted to provide a basic hypothesis of environmental planning on the evidence of hedgerows and riparian linear green patterns, however it is followed by a careful completion with a watchful analysis of the area and surveying.

3.8.3 Selecting New Greenway Segments by Layer

Utilizing the vectorized segments of extracted linear greenway features (7 and 8), a process of selecting the features within these vectorized map layers was undertaken to separate the newest greenway segments from older ones. First, using a 'Select By Layer' function, the extracted features from the 2001 MIVIS image were selected. Various distances were experimented with, as it was desirable to locate areas of potential environmental stress at a slightly larger search area based on the extracted features. Once selected, the results were inverted such that the selected features of the original image were now the features that were greater than the specified distance from any of the extracted features in the original MIVIS image (10). The selected features were then put into a new ESRI shapefile for further analysis, as they had now been determined to be 'new' sections of the extracted features developed within the study timeframe. Ultimately, an integrated RS/GIS multi-source/multi-temporal database was created from the results of the RS land use change detection, ancillary data analysis and the RS feature detection and extraction results.

3.9 Summary

This chapter has outlined the methods used to analyze and classify the study area based on its exposure to potential environmental stresses and its current state of development. First, planning practices in Catania Province and in the Etna park were discussed to set the stage for the necessity of the integrated framework used in the project. Following this, a discussion of the study area was provided including an examination of the morphology of the territory within which the study area is located. Finally, the research design was discussed. This methodology was created to make operational the integrated framework presented and discussed in Chapter 2. Several research techniques were used to obtain the necessary RS imagery and groundbased

GIS data and an integrated RS and GIS process flow was outlined demonstrating the work undertaken. The next chapter presents the results of the research based on this design.

CHAPTER 4
RESULTS AND DISCUSSION

This chapter extends the discussion in Chapter 3 by presenting and discussing the results derived from the previously outlined methods. The results are presented in several sections. First, the outputs of the RS data processing and analysis are discussed. This includes first the temporal change detection and, following the land use analysis, the linear feature extraction is discussed. The second part of the RS analysis is an important step in determining the overall success of identifying areas of potential environmental stress. The RS analysis is also an important step for the semi-automatic methodology for the classification of greenways.

The subsequent feature extraction RS analysis was driven by the results of the combined change detection work and the GIS data analyses, which are then presented and discussed.

Secondary GIS data results are analysed and incorporated with the results of the RS change detection analyses based on the stages of the integrated framework outlined in Chapter 2 and the detailed methods explained in Chapter 3.

4.1 RS Change Detection Data Processing and Analysis

This section builds upon the discussion in Chapters 2 and 3 by presenting the results of the RS analysis stages outlined in the integrated model and the research design. The analyses involved the examination of several indicators of environmental stress detectable in the RS imagery. The analysis steps used in the research include temporal change detection and linear feature identification and extraction. The results of the first part of the RS analysis are now presented in detail. The latter results pertaining to feature extraction are presented following this discussion and the subsequent and land use analysis results.

4.1.1 Change Detection

The change detection analysis and diachronic analysis of the RS images produced results that were successful in terms of differentiating between pre-existing and new development in the Province of Catania and in the Etna park.

As expected, the identified areas of change within the Etna park lie mostly at the immediate zone C and zone D of the regional park. More specifically, change is especially evident within the environmental area termed as the Mascalucia and the districts of Nicolosi and Pedara. Interpretation of the change detection image provided the following results:

1) minimal to no change is noticeable throughout most of the environmental areas, as illustrated by the medium grey tone in the image;

2) some of the notable differences, visually the significantly darker and brighter tones, in the predominately rural areas are attributable to differences in crops and time of year;

3) some of the very bright areas throughout the image may be characteristic of land currently under development (from bare soil to developed land use);

4) the brighter and darker areas at the C and along the D zone represent the areas of greatest change in terms of urban land use development.

Examples of the locations that typify these general observations are denoted and are shown in separate close-up images in Figure 4-1.

Based on these general observations, more detailed conclusions can be stated in terms of the temporal changes at the zone C and D during the time frame under investigation.

This assists in affirming the expectation that this area is, in fact, one of the most rapidly growing areas at the Etna park is likely to be one of the areas most susceptible to potential environmental stresses. This indication of rapid growth is further confirmed with the ground-based land use analysis discussed later in this chapter. Second, it is visually evident from the change detection image that the areas of change form a band along the known edge of the Etna park. In particular, within this band, areas of change are most evident along major greenways (including the waterways). This pattern is consistent with the expected forms of ribbon development predicted in the research project.

Total areas and locations of change are further discussed in the following two sections.

Green Corridors

Medium Gray = No Change Darker and Brighter Tones = Change

Area 1 – Medium gray tone denotes no change in urban regions.

Area 2 – Some change, brighter areas, in rural regions may be due to crop rotation.

Area 3 – Some very bright tones may be attributable to land currently under development.

Area 4 – Pockets of generally darker and brighter areas denote change.

Change Detection, semi-automatic technique and classification.

A semi-automatic classification was applied to the change detection results to provide a homogeneous representation of the green infrastructure and changes within the study area.

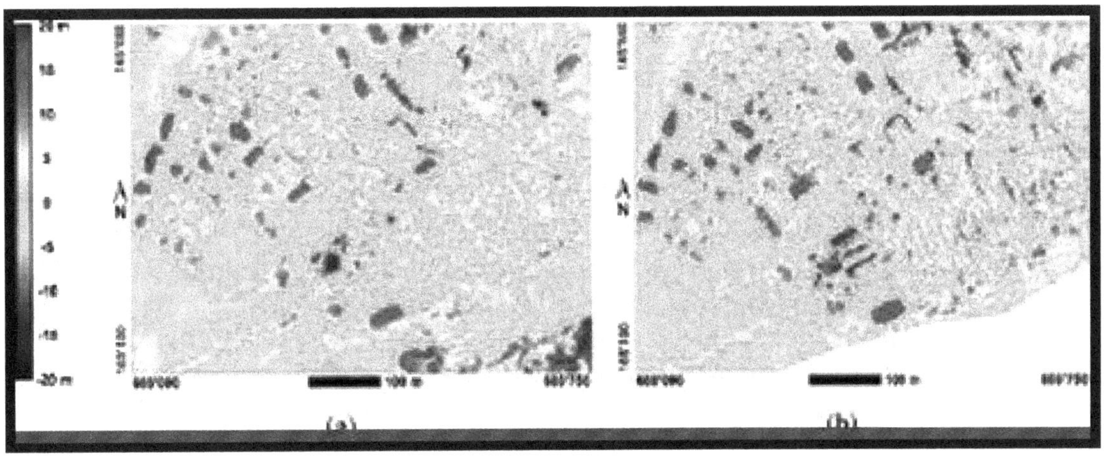

FIGURE 4-1-1: AREAS OF NOTED CHANGE DETECTION BETWEEN 1995 AND 2001

The production of the change detection results in this format allowed them to be integrated with GIS analysis. The total calculated area of change, excluding the areas of no data (no concurrent satellite coverage), was 266.7 square kilometres (26,676 hectares) while the total area of no change was 105,8 square kilometres (10,580 hectares). The total area of both provinces, less the areas of no data, is 132,4 square kilometres (13,2482 hectares). This represents a percentage change of 20.1%.

4.1.2 Change Detection in a GIS

Utilizing the results of the change detection, the image was subdivided to determine the total amount of change in each *zone*. Once divided, the total area of change per *zone* was calculated and then ranked in order from greatest to least. The *zone* with the most change at 29.37 percent (such as Mascalucia, Pedara and Nicolosi in the Catania Province). This represented a total area of change of 35.41 square kilometres (35,41 hectares) out of 124.3 square kilometres (12,430 hectares). This represents a significant amount of detected land use change in these provinces. This presents the results in total areas and percentages of land use change for all the *zone* and it displays the change detection zone rankings based on percentage change for the Provinces adjacent to Catania Province.

These results are supportive of the overall conjecture of this research, namely that a great amount of land use change is occurring along the zone C and D, and in particular, along the main roads. Based on these results, and the land use indicator analysis, the Ecological Corridors is confirmed as an appropriate location for linear feature detection and extraction.

In general, these results can be applied to the stages of development outlined in Chapter 2. Specifically, the *zones* that have experienced the largest percentage of detected change have the greatest potential for environmental stress as they are evolving quickly from zone C to zone D. If this movement is occurring too quickly, these areas are decidedly suspect in terms of localized rapid land use evolution and potential for consequent environmental stress. Supportive evidence that this is occurring are the relatively short time frame being investigated, and the large percentage of detected change, up to 29% in some zones.

4.1.3 Greenway Buffer and RS Change Detection Analysis

The analysis of the greenway buffer and the RS change detection proximities in the provinces of Catania and in the Etna park revealed some important correlations in terms of substantiating one of the conjectures of this research project. By applying a 100 metre buffer to the existing greenway network and selecting areas of noted change within this buffer, a total area of change (14,967 hectares) was detected within 100 metres of existing main cultural roads and itineraries along the Etna Park. This represents approximately 10% of all change noted within the combined provinces of Pedara, Nicolosi and Mascalucia. Increasing the buffer to 200 metres raised the total detected change, approximately 25% of the total detected change (figure 4.2). This total represents a significant proportion of the change within a narrow buffer adjacent to existing cultural itineraries and in particular greenways.

Finally, by increasing the greenway network buffer further to 500 metres, the total detected change within this buffer is (19,498 hectares). This represents 59% of the total detected change of the Catania province. These results are notable in terms of providing a conclusive link between the results of the RS change detection and existing green infrastructure. They also serve to confirm further that a significant portion of new urban growth is occurring in close proximity to main cultural roadways of the green infrastructure.

All spatial data included in my project are also analyzed using the following GISCloud tools:

Area

Buffer analysis

Layer Comparison analysis

I perform Analysis only with layers added from a open data source (e.g. PostGIS, GRASS, etc.).

Area analysis calculates the surface area of the selected objects (e.g. the greenways). To perform Area analysis, first I use the Select tool to select the objects.

Buffer analysis creates a zone around a line or polygon. I identify whether there are objects of interest inside the zone or outside the zone (e.g. the greenways) and make conclusions based on this analysis. After selecting Buffer analysis from the menu I need to set following options:

Analysis name

The layer to perform the analysis on

Define the buffer distance and the unit of measure

Choose whether I want to group (merge) the buffers

Define the symbology of the buffer polygon

See image below for an example of grouped geometry buffer on a point layer.

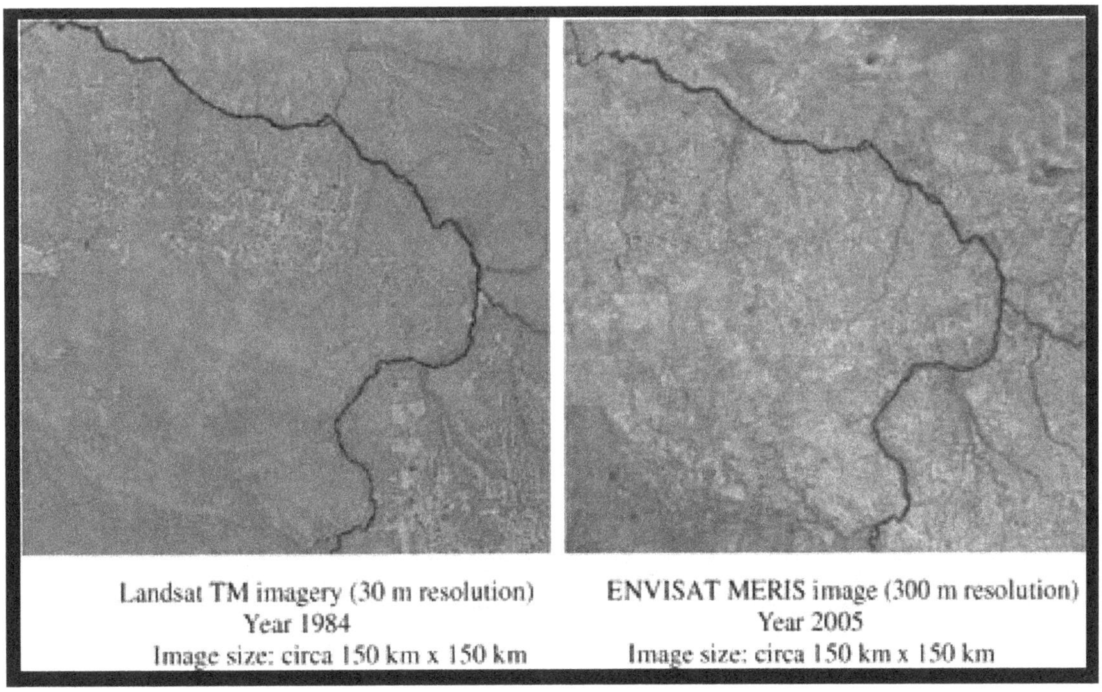

FIGURE 4-1-2: A PARTICULAR OF DETECTED CHANGE WITHIN 200 METRES OF EXISTING GREENWAYS

Layer Comparison outputs area calculations for the intersecting portions of different polygons (e.g. the greenway network)

4.2 GIS Data Analysis and Processing

Prior to discussing the integration of RS and GIS data, the results of the GIS data and land use analysis are presented.

4.2.1 Temporal Land Use Change and Coverage Intersection

The analysis of the different indictors of potential environmental stress in the Catania province produced interesting, yet expected results. it is the district with the highest potential for environmental stress based on the indicators examined (Figure 4-3).

This result is consistent with the general expectation that locations adjacent to recent greenway development show the greatest propensity for rapid land use change and hence potential stress.

One factor that might explain the relatively low score for Nicolosi is that it has an elongated shape from east to west and it is narrow from north to south. Since "Via del Convento" mainway runs north to south through these zones, its growth inducing effects are likely to have been less substantial in the Etna park in Nicolosi district than in either Pedara or Mascalucia.

4.2.2 RS Change Detection and GIS Data Integration and land use analysis

The separate examination of the results of the change detection and the indicator analyses, outlined individually above, generally indicate similar results in terms of areas of change and thus areas of potential environmental stresses. However, alone, neither is as persuasive as when the two methods are combined for environmental stress detection. The change detection results, ranked based on total area of detected change, were added to the

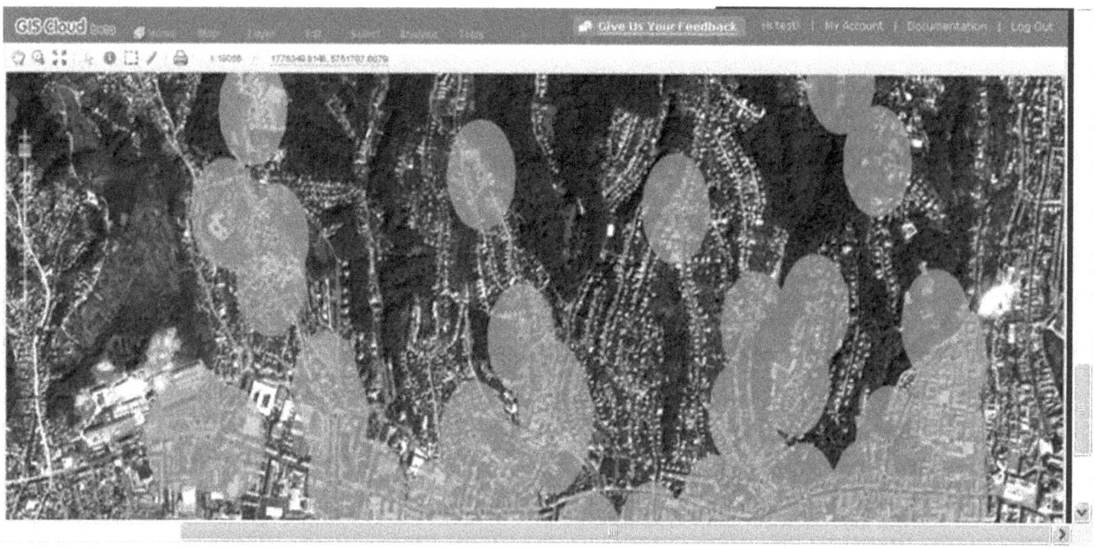

FIGURE 4-2: DETECTED BUFFER OF EXISTING GREENWAYS

FIGURE 4-3: EXAMPLE OF POTENTIAL ENVIRONMENTAL STRESS (BASED ON MIVIS ANALYSIS)

overall rank sum of the different GIS-based indicators. This provided a new overall rank summation. This area has been also studied in the framework of the "Carta della Natura" Project which has led to the "CORINE habitat mapping" based on the criteria exposed in the UE CORINE Biotipes Project. In particular, a landscape description of humid area in the etna park and a detailed lend cover/vegetation map (one to fourth level of detail CORINE) have been explored. The thematic maps produced at 1.10000 scale shows a high diversity of natural and seminatural categories. Landscape pattern is characterised by a dominance of forested and seminatural areas in zone A and B, analysis oh habitat (Direttiva 92/43/CEE) and cultural landscape inside the analysed areas is also analysed. A example of the classification is illustrated in Figure 4-4.

In general, the overall rankings changed little and in fact, the addition of the RS change detection rankings served to confirm further the results of the GIS-based indictor analysis.

4.2.3 Land Use Classification and Localization of Detected Stress Areas

Based on these integrated ranking results, it is evident that the Ecological Corridors (usually narrow structures of vegetation) represent unmistakable region of significant environmental sources and thus has increased potential for environmental stresses above and beyond other regions within the Etna park.

In general, the results of the integration of data derived from both the RS and GIS analyses, has, at the district level, divided the Catania Province into areas of varied susceptibility to environmental stresses. This classification system can be extrapolated into the stages of development outlined in Chapter 2. Specifically, the zones of greatest potential for

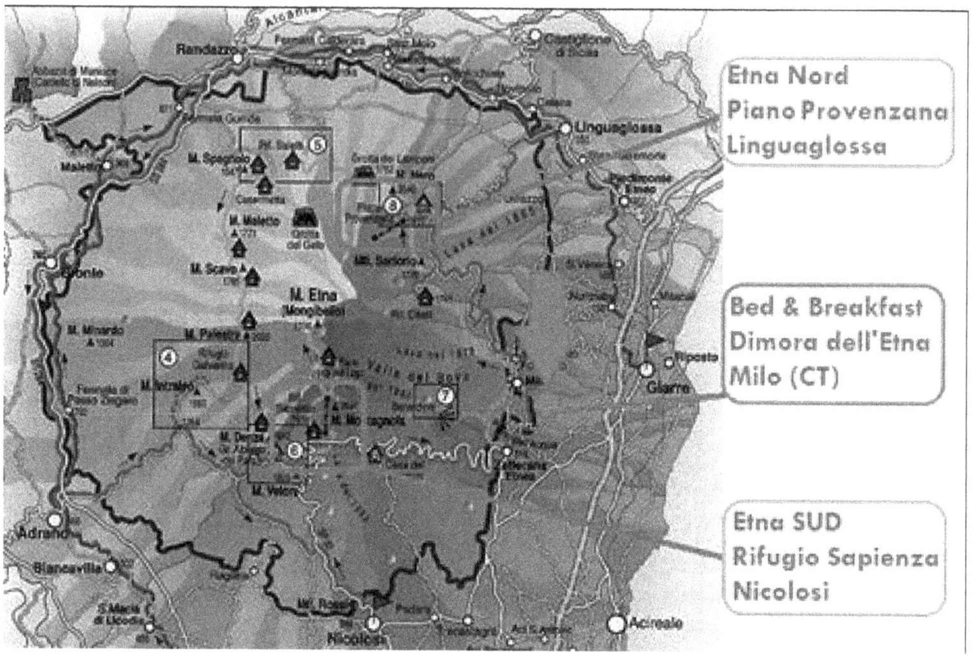

FIGURE 4-4: LAND USE CLASSIFICATION AND IDENTIFICATION OF CULTURAL LANDSCAPE AND GREENWAYS

environmental stress based on these ranking are most likely to be located in zone C and D in terms of built-up land. These districts are likely to be characteristic of these stages of development in terms of the RS change detection analysis.

This type of generalized stage classification is possible because the RS change detection were selectively included in this analysis given their expected potential to indicate change from rural to urban land use and land cover mapping (Kavzoglu, T., Colkesen, I.. 2013)

4.3 Linear Feature Extraction

The next step in the overall operational framework focuses on the detection and extraction of linear features from temporal imagery in an attempt to locate new green infrastructure development as potential indicators of growth and of higher potential environmental stress. Because the Ecological Corridors was revealed, through earlier analysis, to be the area in Pedara that has experienced some of the largest actual and implied environmental growth, this sub-region was targeted for feature extraction analysis. The results of this process are now presented.

Based on preliminary results, it was evident that the maximum likelihood algorithm produced good results in terms of quality of linear edge detection and number of edges located. Further, because the Canny algorithm was specifically designed as an edge detector and has been described as one of the 'most powerful' edge detection method available (The MathWorks Inc., 2000 and John A. Richards, Xiuping Jia, 2013), it could also deemed the good option for continued analysis in terms of linear feature detection and extraction.

The threshold value used with the maximum likelihood algorithm produced varying ranges of extracted edges. Ultimately, a threshold of 0.2 was used for the MIVIS image and a threshold of 0.1 was used for the lower resolution Landsat images and Spot PAN

Image to ensure the capture of all desirable training areas, edge detection and in particular, any known or potential greenway segments.

Natural and seminatural mapping was the key issue for planning and conservation and for defining monitoring programs. The individuation and classification approach has provided promising results in the implementation of new green infrastructures.

4.3.1 Edge Detection with Landsat TM (Tematic Mapper) Imagery

The Landsat TM from 1995 performed, in general, inadequately in terms of its use for effective linear feature extraction. A visual inspection of the results quickly reveals that the low spatial resolution of the image did not allow for an effective detection and extraction of many discernable features, including greenways.

Even though several pre-detection image enhancements had been performed, as discussed in Chapter 3, the results for the Landsat image in this scene were inferior when compared to results achieved with other higher resolution imagery.

Due to the unsatisfactory results produced with the Landsat image, it was removed from further examination in terms of attempting to detect the development of new roads in the Etna Park. Unfortunately, this left only a one-year time frame for investigation since the remaining available images were captured in 1995 and 2001 respectively.

Nevertheless, given that one of the objectives of this research is to capture newly developed green infrastructure at the rapidly expanding zone C and B, it is possible that new green infrastructure were developed in a time period of less than a one year.

4.3.2 Edge Detection with MIVIS Multispectral Imagery

The MIVIS multispectral image from 2001 produced the best relative results in terms of edge detection and linear feature extraction (Figures 3-9). With the highest spatial resolution of the images the MIVIS image detection results are visually superior. Nonetheless, similar problems were experienced to those of the Landsat image, such as the detection of non-linear and non-road features. However, these were not as problematic as with the Landsat image. Also, many known features, including green infrastructure, were also apparent as successfully extracted items. Relative to the other two images, the MIVIS Multispectral image produced the most extracted features, also due to its higher spatial resolution.

4.3.3 Edge Detection with Spot PAN Imagery

The results for the Spot PAN image feature extraction can be placed in-between those of the successful MIVIS image and the Landsat image. This can once again be attributed to the spatial resolution, as the Spot PAN image has a slightly lower resolution than the MIVIS image. It is clear from the image segment that the edge detector, when utilized with the lowest threshold, has the ability to detect every edge in the image notwithstanding varying image intensities. In terms of the linear features, given the relatively low spatial resolution of the images under investigation, the algorithm was most effective at detecting larger greenways within the study area. While detectable, secondary greenways were visually difficult to distinguish from other types of edges detected within the imagery.

4.3.4 Combined Edge Detection Results

Common amongst all of the imagery examined is the fact that the number of edges detected and extracted is more prolific in areas of the ecological corridors and greenways as compared to the relatively more natural areas. This phenomenon is particularly prominent along the major arterial greenways and is evident immediately adjacent to Convento way, which bisects the Nicolosi area. Moving from south to north along the greenway, a pattern of development is apparent adjacent to the way relative to more outlying locations.

Aside from these edges, it is also evident in Figure 3-9, that many non-linear and non-road man-made and natural features have also been detected and extracted by the semi-automatic approach (chapter 3). This pattern of extraction supports one of the conjectures of this research project in that the ability to extract linear features from satellite imagery is greatly affected by the location of the study area.

Specifically, an urban study area tends to produce extraction results that are more disorderly and less successful relative to a rural study area. This is due to the density of features, manmade and natural, in an urban area with spectrally unique image values. This proliferation of unique values in close proximity forces the detection and extraction of many edges in the urban and environmental scene.

Following the extraction of the edges, the SPOT, the Landsat and the MIVIS images were written to TIF image format and then imported into ArcGis. Using the 'gridline' command, they were then automatically converted to ARC coverages.

Edge detection is a fundamental tool in image processing. Several edge detectors have been proposed in literature for enhancing and detecting edges in images. Image Edge detection significantly reduces the amount of data and filters out useless information, while preserving the important structural properties in an image. Edge basically occurs where there is significant change in intensity. The principle of the algorithm used is to increase the difference between those pixels where intensity values change significantly. So by using this concept, detected edges are wider and clear. The given algorithm can be applied to gray scale and monochrome images (Pratibha Sharma, Manoj Diwakar and Niranjan Lal. 2013).

4.4 Further Integrative RS and GIS Analysis of the Ecological Corridors

This section provides an examination of the integration between the features extracted from the RS imagery and the further temporal analysis of these features in a multisource GIS and GISCloud.

The results of the conversion from GRIDs to ArcGIS coverages further reveals the over abundance of edges that were detected and extracted. In particular, the Landsat image is inundated with edges that are clearly not representative of greenways in many cases. To improve these conditions, some manual editing of both ArcGIS coverages was necessary to remove some of the more obvious non-greenways features. However, limited manual removal of lines was performed to ensure that potential minor segments were not mistakenly deleted. Also, manual checking was completed to verify that any discrepancies between the extracted lines from the two images were minimal and that the extracted lines were coincidental to allow for the ensuing overlay process. Discrepancies were potentially caused by several possibilities including geo-referencing inconsistencies, edge extraction variability and raster to vector conversion divergences. In general, minimal, and largely insignificant inconsistencies were visible between the two images.

One other source of extracted feature discrepancy represented the largest source of inconsistency between the extracted lines. This divergence was a result of the difference in image spatial resolution. Some of the extracted features are misaligned in the two time periods by up to a maximum of approximately 50 metres. This number is consistent with the differences between the two spatial resolutions, eight for the Landsat versus 24 for the Spot PAN, and the range of variation potential that was possible for the same features in each image. This range is evident where the pink (Landsat) extracted arcs (Figure 4-5) do not match up exactly with the blue (Spot) extracted arcs.

To account for this variability in spatial resolutions and subsequent positional accuracy dissimilarities in the feature selection by theme stage, the next and final step of this research uses a wide selection distance to ensure only potential new line segments are selected.

4.4.1 New Feature Segment Selection by Layer

Utilizing the results of the 'gridline' conversion along with some subsequent manual coverage editing, a selection of temporally extracted features was undertaken. Using a 'Select By Layer' procedure, features from the Spot image were selected that were within the 2001 MIVIS image. Distances ranging from 5 to 50 metres were input in an iterative process of determining the most successful at returning the highest number of potential new features. Ultimately, it was determined that a median distance of 25 metres was most efficient at capturing more established new features and eliminating minor edge variances that were being captured at a shorter buffer distance. Once the selection by digital layer was completed, the resulting selection was inverted such that the selected items were now the features located at the specified minimum distance from any feature extracted from the 2001 image. While it is quickly apparent that many of these selected features are not greenways, the possibility that some of them are new green infrastructure is high given that many of the arcs are located in-between major arterial greenways. This would suggest that some of these new lines might be greenways that would have been recently constructed to link major cultural sites and heritage. Alternatively, these new lines may simply be new extensions to existing greenways as growth occurs outwards into new areas of development. The only method of determining whether or not these new lines are actually greenways was through manual visual inspection of each arc overlaid with the Spot and MIVIS imagery to determine actual feature status. These research was followed by surveying activities (laser scanning surveying) and by a watchful analysis of the area.

Latest release of SmartPlant 3D further solidifies its position as the preferred automated 3D design solution for executing large, complex capital projects. SmartPlant 3D delivers new technology advances in interoperability and design reuse to help you:

- Significantly reduce the time needed to get projects up and running
- Compress the time schedule required for their completion
- Materially reduce the capital costs for design

Reference 3D offers interoperabilty with many different 3D models, making Smart 3D the GIS. Smart3D can reference and perform clash reporting and provide critical design references on multiple models, with multiple formats, simultaneously.

4.4.2 New Greenway Determination Results

Careful examination of these new lines reveals the potential for some preliminary conclusions in terms of whether extracted features are actually representative of new greenways development or not. However, given the inability to state with a high degree of certainty whether these features are green infrastructure segments, conclusions are speculative at best. For example, some segments branching out along the Convento Greenway may indeed be new greenways supporting the rapid environmental sprawl occurring along the green infrastructure (Figure 4-5).

Unfortunately, given the generally low spatial resolution of all of the imagery under examination, it is impossible to conclude that any of these segments are indeed new greenways.

Therefore, in terms of the results presented above, while the outcome in this context is less than conclusive, it can be stated that the operational process outlined here could potentially be utilized in other more conducive situations and alternative study sites to detect and effectively extract features from satellite imagery. Specifically, an entirely rural study area would be more likely to allow for the generation of improved results when utilizing this operational process, as the number of extracted image features would be greatly reduced, as would be the intensity and number of edges.

On the other hand the classification approach explored the potential tool of the semi-automatic approach in order to classify the green infrastructure. We could be able to identify six main categories (main ways, footpath, "ciclopedonali", "cicloturistiche", "Regie Trazzere" and raiway) with the support of the NIR band and the analysis of the NDVI and diffrent spectral firms (chapter 3).

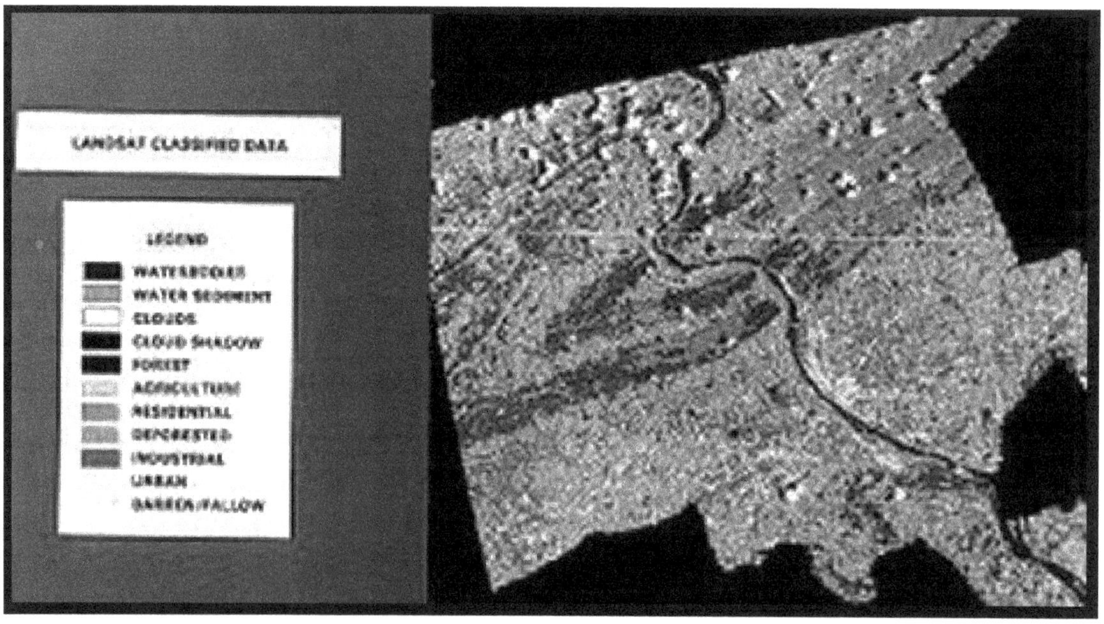

FIGURE 4-5: THE PINK EXTRACTED ARCS (FALSE COLOUR, LANDSAT)

4.6 Summary

This chapter has described the results derived from the implementation of the integrated framework outlined in Chapter 2 and further discussed in the context of an operational methodology in Chapter 3. Starting with the RS change detection analysis, the results of each stage of the methodology were presented and reviewed. This was extended to include the change detection and green proximity analysis, the land use-based GIS indicator analysis, the localization of areas of potential environmental stress and the linear feature detection and extraction with the semi-authomatic approach which was suggested to become ecological corridor or greenways. The final product was a multi-source and multi-temporal GIS database that had the capability to locate areas of cultural landscape and environmental values in the Etna Park. Based on these results, Chapter 5 presents the conclusions of the project. Also, a review of the overall effectiveness of the process is discussed along with some recommendations for further research.

CHAPTER 5
CONCLUSIONS AND RECOMMENDATIONS

This chapter provides a synopsis of the central arguments presented in this book and outlines the conclusions derived from the results obtained. Also, suggestions for growth management policies with respect to improved policy formation and planning practice are presented. Finally, some recommendations for future research are discussed as well as practical applications and implications for the use of RS, GIS and geomatic technologies to achieve sustainable environmental management (SEM) in parks and cultural landscape.

5.1 The Synopsis

This book delivers information and technology skills to people around the world. It provide a comprehensive overview of networking, from fundamentals to novel applications and services. The book emphasizes theoretical concepts and practical application, while providing opportunities for you to gain the skills and hands-on experience needed to design and analyse novel geomatic applications. In particular, the book implements the curriculum of various disciplines involved in cultural and natural heritage conservation.

This book also represents the official supplemental textbook for Under-graduated courses in cultural and natural heritage curriculum for Universities and Research Centres. As a textbook, this book provides a ready reference to explain the integration of Remote Sensing (RS) and Geographic Information System (GIS) for Cultural and Natural Heritage Conservation and Valorisation.

This book emphasizes key topics, terms, and activities and provides many alternate explanations and examples as compared with geomatic courses and programmes. You can use the book as a reference book in geomatic programmes and as a main textbooks in courses on cultural and natural heritages to help solidify your understanding of all the mentioned topics.

The main objectives of this book are: (i) to develop to an integrated framework based on the concepts of Sustainable Development (SD), Environmental Mamagement (EM) and Cultural Tourism (CT), for the achievement of Sustainable Environmental Mamagement (SEM) and for the detection of land use change, indictors of potential environmental stresses and identification and classification of green infrastructures in Parks and preserved areas. The Etna Park *has been chosen* to act as an example of *test sites* for novel SEM. The need for a methodology to evaluate cultural landscape conversion is apparent from the current planning and growth management systems in place. Typically, planning decisions are *ad hoc* and reactive in the absence of up-to-date information on cartographic sources. Even with past attempts at long-term planning, the provision of necessary green infrastructure has dragged well behind growth rates in Geographic Information System (GIS) platform.

Urban sprawl and various disastrous human interventions (such as flow variations, excavation works and dike building) are placing immeasurable amounts of stress on natural systems. To address theses stresses, improved sustainable environmental management is necessary. To this end, the integrated framework that is presented and analysed in this book seeks to address some basic needs of SEM through the use of spatial information technologies. Through this integrated and structured framework, areas of potential environmental stress at the different zones are identified and classified so that they can be targeted for proper sustainable growth management.

The need for this integrated approach was made apparent in Chapter 1 and 2 where Remote sensing (RS), GIS, sustainable development and environmental management were first discussed in the context of cultural landscape and heritage. Next the use of Spatial Information Technology (SIT), including Remote Sensing (RS) and GIS and novel implementations were discussed in terms of their uses for improved planning practice in natural systems. An integrated framework was then introduced uniting the concepts of SD and EM to form SEM, with the operational processes created through the use of SIT.

Chapter 3 presented the Research Design developed to achieve the objectives presented in Chapter 1 and discussed in Chapter 2. Also, the planning practices in test sites were addressed. The discussion turned to an analysis of the specific study area and its morphology. The design of the methodology, discussed in the last half of Chapter 3, was created to make operational the integrated framework outlined in Chapter 2. The design is comprised of RS imagery analysis, ground-based GIS data analysis and an integrated analysis combining the RS, GIS source data and GISCloud.

Chapter 4 presented and discussed the integrated framework and analyses results derived from the methods outlined in the research design. The discussion in this chapter was structured to illustrate how the integrated framework was made operational through the methods used in the book. In particular, this chapter demonstrated how potential indicators of environmental stress that are detectible in RS imagery and derivable from ancillary land use analysis can be combined and analyzed within a multi-source, temporal GIS database and GISCloud to facilitate improved SEM and CT.

5.2 Conclusions

Conclusions regarding each portion of the research are presented here including the image analysis, linear feature extraction, the indicators and locations of potential environmental stresses and the individuation and classification of the green infrastructure at the different zones of the Etna Park.

5.2.1 RS Change Detection Analysis Conclusions

Overall, the steps undertaken in the RS imagery analysis produced some interesting and useful results. First, the change detection using image arithmetic produced anticipated results in terms of the determined locations of change. As noted in Chapter 4, the areas of detected change corresponded with the presupposed locations of change in terms of their locations along the periphery of Catania Province. Some further examination and alternative change detection experimentation, such as principal component analysis or image overlay, could potentially reduce errors in the change detection results and semi-automatic method and classification. However, given that the results were as predicted, it is expected that only slight improvements in accuracy could be achieved for this section of the research methods topic. In part, the low spatial resolution of the imagery assisted in facilitating these results by producing a general smoothing that would have otherwise been cluttered by extensive detail in a higher resolution image. This produced a more homogeneous surface for the change detection process to detect and classify larger adjacent areas of natural structures of the green infrastructure. However, a consequence of this was a potential loss of precision in terms of localizing change.

5.2.2 Land use Analysis in GIS Conclusions

The incorporation of the Land use Analysis analysed in the research project also provided useful information in terms of remote sensing analysis (including the *vegetation index and NDVI analysis* based on the NIR of the MIVIS over the frame examined). The results of this analysis was generally as expected such that the *thematic map* showing the greatest amount of change based on the combined indicators were very similar to those selected with the other RS change detection. These results serve to reaffirm the selection of the Ecological Corridors as natural areas able to connect the zones with peculiar naturalistic-environmental values within the Etna park that are rapidly developing along the different zones of the park.

5.2.3 RS and GIS Integration and Localization of Stress Areas

Throughout the research, the benefits of RS and GIS integration were stressed. This integration was proven to provide timely and necessary information about land use change to planners in natural environment and parks. Overall, the integration of the RS change detection results provided increased utility and validity to each other in terms of raising the level of certainty in localizing the regions with the highest potential for environmental stress.

This assisted in the localization of areas of stress that ultimately led to the identification of the Ecological Corridor and green infrastructure as appropriate areas within the Etna park for more detailed analysis of greenway feature extraction.

Overall, the framework presented here represents an important contribution to the future prospects of RS and GIS integration in terms of analysing change for the management of natural and cultural resources with the support of "open data". The localization of areas of stress will help to ensure improved environmental management through the timely detection of areas requiring protection and improved environmental supervision. Protection could potentially come in the form of an environmentally sensitive area designation while improved supervision may involve increased monitoring of development to ensure adherence to environmental protection guidelines. Once these areas are located, modifications to government policy to include future protection and improved growth management, such as region-wide infrastructure planning or integrated waste management systems, will ultimately assist in the provision of SEM throughout the cultural and environmental sources.

Development of a growth management strategy and valorization of cultural resources involves a balanced combination of planned growth, environmental integrity and economic viability. The integrated framework presented in this research has the potential to provide for each of these items. New policies must be implemented to ensure planned growth as opposed to existing random development patterns. However, these policies cannot be devised without adequate information related to the location and form of new development, and this is where the integrated framework can make a substantial contribution to planning policy formation. This analysis can also support the implementation of the tourism market and in particular the CT. Environmental integrity can be achieved through an improved planned growth initiative. We can not think about CT without the environmental integrity of a territory. The need to protect prime cultural landscape and environmentally sensitive areas is important in terms of the economy of a region and also to the quality of life of its residents. Natural resource protection and conservation must become a primary objective. The integrated framework can also serve to facilitate improved planning practice in terms of early detection of areas of environmental stress that may require an environmentally sensitive area protection policy.

Finally, economic viability is something that all regions, and countries strive to achieve. As noted in Chapter 2, the desire to achieve economic growth and profit has often come at the opportunity cost of negative impacts on the natural environment. It is hoped that the integrated framework presented in this research might be able to assist policy formation in regards to improved land use planning through the provision of timely land use change information (diachronic and sinchronic analysis). This will potentially ensure environmental sustainability and land use compatibility while at the same time not hindering economic prosperity.

5.2.4 Feature Extraction and New Greenway Determination Conclusions

Russ (1995), Fasel and Khan K. A (2012) pointed to the following as the two major problems with edge detection:

1) it cannot by itself complete the segmentation of the image because it has to be given each new starting point and cannot determine whether there are more outlines to be followed; and

2) the same edge-defining criteria used for following edges can be applied more easily by processing the entire image and then thresholding … [however] this produces a line of pixels that may be broken and incomplete.

These two points are evident in the results achieved in this research. In particular, the automatic processing of the Ecological Corridors and Greenways resulted in broken pixels and incomplete data, not to mention an overabundance of detected and extracted features.

This analysis has provided a general picture of potential new greenway development in the cultural landscape. Unfortunately, the results are disorderly and it is difficult to prove with any high level of confidence that the extracted features can be accurately identified as new greenways. In order for a more detailed picture to be developed, the extraction process would need to be examined in further detail to determine if a means of more accurate greenway identification could be realized.

This further examination could potentially involve a three pronged approach including a reevaluation of the commercial software and algorithms used in the research; the potential use of a pre-existing or newly developed linear feature extraction stand-alone application to the images under investigation; and the substitution of the current images with new, higher resolution images into the existing integrated framework for renewed analysis.

The use of alternative commercial software with enhanced feature extraction capabilities could provide improved results in terms of the quality and quantity of the extracted features. Within these alternative packages, algorithms other than the Canny algorithm may be available that might be more successful in extracting man-made features such as greenways. Related to this, the use of a pre-existing feature extraction application may also improve results. For example, the LSB-Snakes application, designed by Gruen and Li (2002) and discussed in Chapter 2 and futher implementations (Fasel Q. and Khan K. A, 2012) may be a viable alternative. To date, the use of LSBSnakes has been successful in natural scenes, but has yet to be fully tested in an urban setting.

The substitution of higher resolution images is discussed further below in the recommendations and implementation section (5.5).

Since one of the main objectives of this research was to provide a semi-automatic means for the extraction of linear features, the overall results of this portion of the analysis are disappointing. However, due to a general inability of an edge detection algorithm to distinguish successfully between extracted features, little can be done at this point to improve the results. Consequently, based on the available edge detection technology and the complexity of the study area, the results can be generally considered successful as potential new greenway features were indeed extracted and localized. These new greenways segments can further be used to identify specific areas of zones with peculiar naturalistic-environmental valences and expansion into new, previously undetected, areas of land use development.

5.3 Growth Management Policy

The failure to manage the environment and to produce sustainable development is evident in all countries. Because environmental management and development are not separate challenges, environmental stresses and patterns of economic development are closely linked (WCED). Based on the results of research projects, it has been demonstrated that there are many facets of environmental planning that may be better facilitated through the use of spatial information and geomatic technologies. Specifically, the localization of areas of potential environmental stress through the integrated RS change detection and GIS land use indicators procedure, will allow local planning authorities to target problem areas for improved green infrastructure provision and to make possible sustainable environmental management.

In terms of further facilitating growth management policies, the *semi-automatic* results based on the integrated RS and GIS indicator analyses will provide a particularly effective means to determine the zones with peculiar naturalistic-environmental valences for the valorization of the heritage including cultural landscape. Thus, improved, more environmentally sustainable policies can be created and applied to the *semi-authomatic approach*. As noted above, these approaches may also be used to direct general growth management policy in terms of applying stages of development to the provinces and districts adjacent to the Etna Park.

Specific environmentally sustainable policies must include some or all of the following requirements for future development:

1) A region wide land use and green infrastructure development plan;

2) Protection and environmental enhancement of environmental and other natural resources;

3) Providing an integrated regional green infrastructure system among natural areas and parks in Sicily;

4) Managing and controlling various disastrous human interventions (such as flow variations, excavation works and dike building) to ensure service and to minimize negative environmental impacts; and

5) Monitoring change to appropriately adjust policies and actions. (adapted from Regional Commission in Sicily, P.O.R. 2000-2006 and further bimplementations)

6) Providing an integrated European green infrastructure system among natural areas, cultural landscape and parks in Europe (MEDDOC - P.I.C. - INTERREG);

The creation and implementation of policies with these requirements in mind will only be possible if an European approach to planning within the broader natural areas, cultural landscape and parks is undertaken through partnerships between the local authorities and European authorities. The integrated framework presented in this project seeks to assist in the creation and implementation of such policies by providing more timely and accurate information pertaining to the location and morphology of natural and environmental sources at the district level within the Regional Etna park and at the European level. This information is a fundamental input to the creation of accurate and informed policy to guide growth and provide protection of the rural, natural environments and cultural landscape in general. These data are also fundamental in the monitoring of change to allow for appropriate adjustments to existing policies and for providing a starting point for action in terms of rectifying improper and environmentally unsustainable sprawl and various dike buildings. These data are also essential in the development strategies in the field of CT in order to allow for appropriate adjustments to existing policies and for providing a starting point for action in terms economic development.

5.4 Objective Conclusions

Overall, the integrated framework developed in the research project, based on the concepts of SEM, has been shown to make operational an effective process of environmental stress detection through the integration of RS and GIS spatial information technologies.

Specifically, RS change detection analysis has been shown to allow temporal land use classification in terms of change versus non-change areas in the Province of Catania within the Etna park. Additionally, the GIS land use indictor analysis has also been shown in these zones in terms of land use change and potential environmental stress.

The temporal extraction of greenway features from RS imagery and GISCloud is one of specific objectives of this research that has not been completely fulfilled. Generally, the process utilized was effective at detecting and extracting features. Some recommendations on how these results might be improved are discussed below.

5.5 Recommendations and Implementation

The principal recommendation that comes out of this research is the need to transfer the methods outlined here to other scales of research. Specially, if data were available at, for example, the property or plan of subdivision levels, the process of analyzing the groundbased indicators could be moved to a larger scale to provide a finer level of detail in terms of accurately detecting areas of potential environmental stress. By acquiring this level of data (including open data), detailed population characteristics could be detected and analyzed. Unfortunately, the collection of this level and type of data is currently very problematic, especially in natural areas and parks.

The use of higher resolution satellite imagery would allow for a more discrete analysis of the existing greenway network. As mentioned in Chapter 2, high-resolution satellites such as IKONOS and QuickBird, whose imagery are becoming readily available to researchers, will allow for the detection and analysis of detailed greenway characteristics. However, as the green infrastructure become more distinct in these high-resolution images, other features also become more visually and spectrally unique relative to their adjacent land uses. This means that rooftops, building footprints and other detailed features will appear as sharp edges in terms of feature extraction. Hence, the use of these images in an environmental setting could become even more problematic than their lower resolution ancestors.

To overcome this problem, the algorithms utilized in the detection and extraction of edges need to be more intelligent than the available automatic and semi-automatic algorithms. As discussed in Chapter 2, several on-going research projects are attempting to automate further the process of linear feature extraction. Many of these projects are operating in exclusively natural areas and only a few have ventured into the urban fringe and/or an entirely urban and environmental setting. Until edge detection algorithm technology becomes 'smart' enough to discriminate between adjacent land uses that are spectrally different, migrating the processes used in natural scenes will not be possible.

In terms of implementation, given the complexity of the processes utilized in this research project, some potential requirements are recommended. First, and particularly for implementation in natural environment, a less complex process must be established. Fewer steps would be beneficial in terms of implementation by streamlining the process. This could be partially achieved by limiting the number and types of software packages necessary to produce viable results. The use of ENVI, and ESRI softwares in this research was required in an attempt to generate the best possible results in terms of land use change detection and feature extraction. Unfortunately, the result is a complex process that involves a relatively large amount of data conversion across expensive software platforms and different spatial data structures.

However, the development of such an application will not be possible unless all analyses done in both ESRI and ENVI are removed from the process. This would result in a large loss of RS imagery analysis and utility and is therefore not a recommended option of simplification if this framework is to be implemented as designed. While the difficulties of implementation of these methods increases with complexity, the framework as presented represents a favourable methodology for integrated RS and GIS analyses to detect areas of potential environmental stress.

Despite the less than satisfactory results of the linear feature extraction portion of the research, other results at the district level provided a substantial contribution through the identification of trigger points of environmental stress and subsequent policy development. An important benefit of implementation can be realized when this integrated framework is used to address policies related to growth and provide value for decision-making. By providing utility to the process of detecting and rectifying areas of environmental stress, the framework presented in the research has the potential to bring SEM practice into the planning process within the Etna Park.

In the field of cultural preservation and valorization, greenways can therefore implement a system of innovative didactic methodologies in geomatics (Lo Tauro, A. 2012). This network can stimulate innovative educational activities related to the resources of the territory and to new innovative applications under a pluridisciplinary framework. The aim is the implementation of innovative approaches for the analisys of heritage and cultural landscape under Educational Programmes.

Remote Mobile Investigators (RMI - equipped like Mobile Mapping System), Satellite Positioning (GPS) and INS (Inertial Positioning System), integrated with GNSS (Global Navigation Satellite System) including EGNOS (the European system) and later on with GALILEO represented the novel strategies to be used in the collection of such a data. "Selection criteria from the point of view of planning practice should not only include aspects of rendering quality, but also aspects of integration with standard applications at object and landscape scale, such as CAD, GISCloud, DBMS, and terrain modelling, novel wireless applications and open data". "A very simple way of landscape visualisation is the draping of airphotos or high-resolution satellite images over DTMs" (Muhar, 1999 and Lo Tauro, 2010). I also include geomatics in order to implement novel educational appoaches (on-line and face-to-face) for the analysis of digital modelling and digital terrain models (DTMs) for cultural landscape and heritage including areas of D.O.P and I.G.P. production, protected vegetation, green monuments, agricultural parks, protected areas (such as SIC and ZPS), or cultural buildings. Also the attribute data and study area photographs were used to assist with image analysis and interpretation. Some tests were carried out in the Etna Park and were necessary for testing the new instruments reliability and its survey potentiality with GPS positioning technique (Real Time Kinematic), GNSS, wireless applications, novel sensor fusion and underwater technologies. The tests are still under implementation. Change detection analysis, discussed above, is better facilitated with this information by helping to improve image interpretation and to assist in the identification of certain image features in the field of "in situ" analysis. This approach can be estended to the analysis of other European parks. In the case-study was applied the cultural and natural heritage analysis in relation to the location of the greenway network or natural areas able to connect zones with peculiar cultural sources. In the field of cultural conservation and valorization (L.137/02), greenways can therefore create a system of a cultural network for the implementation of European cultural tourism. This network can stimulate conservation strategies related to the resources of the territory or, in other words, to the man-made and natural sources and the implementation of cultural tourism also with the support of ITC technologies under European Quality Standards (such as Minerva and Michael Programme, Lo Tauro, 2003, 2005b, and 2012). All good research poses as many questions as it answers. This research project has done both by answering questions pertaining to the location of areas of environmental stress at the zones C and D of the Etna park, while posing new questions about how to improve the detection and extraction of greenways features from satellite imagery, GIS and "open data".

The European Commission held a public consultation on open access to research data on July 2 in Brussels inviting statements from researchers, industry, funders, IT and data centre professionals, publishers and libraries. The inputs of these stakeholders will play some role in revising the Commission's policy and are particularly important for the ongoing negotiations on the next big EU research programme Horizon 2020, where about 25-30 billion Euros would be available for academic research. Five questions formed the basis of the discussion:

- How we can define research data and what types of research data should be open?
- When and how does openness need to be limited?
- How should the issue of data re-use be addressed?
- Where should research data be stored and made accessible?
- How can we enhance "data awareness" and a "culture of sharing"?

Contributions from the researchers' perspective emphasised that data, metadata and other documentation should be made available in order to be able to replicate the results of a research article and more data available means more scrutiny and getting more value out of the data. Furthermore, there is a need for pre-registration of studies in order to understand the full picture of a research field where e.g. negative results in the biomedical sciences (as well as many other fields) are not published. Then, this is also a need to have binding mechanisms e.g. required data management plans, better linkage between the research data and scientific publication with enforcement of data availability by journals, but also sustainable plans for making data available, where open access to data is formally a part of the research budget.

Searching and finding research data should be also made easier, as open access to data does not necessarily mean accessible data. There was also an emphasis that every contributor should be known and acknowledged and there is a need of establishing cultures around data sharing in different disciplines and "augmenting the scientific infrastructure to be technical, social and participatory" (Salvatore Mele, CERN, See more at: http://openeconomics.net/2013/07/09/open-access-to-research-data/#sthash.USv9VDdt.dpuf)

The answers provided will assist environmental planning through the application of spatial information technologies to provide the basis for a better understanding of change detection in environmental zones with peculiar naturalistic valences.

GLOSSARY

CRoRS Center for Research on Remote Sensing

CELADE American Demographic Centre administrative boundary – equivalent to a province

CIDA Canadian International Development Agency

CT Cultural Tourism

ECNC European Centrer for Nature Conservation

EM Environmental Management

GDRC The Global Development Research Center

GIS Geographic Information System

GMES Global Monitoring for Environment and Security

GPS Global Positioning System

GNSS Global Navigation Satellite System

ESRI Environmental Systems Research Institute

IFTS "Istruzione e Formazione Tecnica Superiore"

IT Information Technology

MIVIS Multispectral Infrared and Visible Imaging System

MMS Mobile Mapping System

PCI PCI Geomatics – Producer of EASI/PACE Imaging Analysis Software

PPP Precise Point Positioning

RS Remote Sensing (can include satellite remote sensing or RS)

SD Sustainable Development

SIT Spatial Information Technology

SEM Sustainable Environmental Management

TM Thematic Mapper (Landsat TM)

EM Urban Environmental Management

REFERENCES

AA. VV. (1993), CORINE Land Cover. Technical guide. ECSC- EEC – EAEC, Brussels, LUXEMBOURG

A.A.V.V. (2000) *I paesaggi umani*, Touring Club Italiano, Milano

Achermann, Friedrich. (1999). "Airborne Laser Scanning – Present Status and Future Expectations." *ISPRS Journal of Photogrammetry & Remote Sensing*. Vol. 54, pp. 64-67.

Amissah-Arthur, Abigail, Bernard Mougenot, and Maud Loireau. (2000) "Assessing farmland dynamics and land degradation on Sahelian landscapes using remotely sensed and socio-economic data." *International Journal of Geographical Information Science*. Vol. 14, No. 6, pg. 583-599.

Adepeju, A. and Oluwole, O. (2012) Sustainability Issues in Urban Streetscapes: A Case Study of Abeokuta, Nigeria. ICSDEC 2012: pp. 252-259.

APAT, (2004), Carta della Natura alla scala 1:50.000. Metodologia di realizzazione, Manuali e Linee Guida 20/2004 APAT.

Arturo Olvera-López, J. Ariel Carrasco-Ochoa, J. Francisco Martínez-Trinidad, J.. (2010) A new fast prototype selection method based on clustering. *Pattern Analysis and Applications* 13:2, pp 131-14, Online publication date: 1-May-2010.

Askew, Dr. Marc. (2002). "Landscapes of the Urban Fringe. A Study of Social, Physical, and Economic Changes at the Regional and Local Levels." *Final Seminar: Methodology for the Implementation of Urban Planning*. JICA Joint Research Program, Department of Urban and Regional Planning, Faculty of Architecture, Chulalongkorn University, Bangkok, Thailand, pg. III 2-1 – III 2-45.

Axelsson, P. (2000) *DEM generation from laser scanner data using TIN models*. The International Archives of the Photogrammetry and Remote Sensing, Amsterdam, The Netherlands, Vol. XXXIII, Part B4/1, pp. 110-117.

Baltsavias, Emmanuel P. (2000). "Digital Ortho-images – a Powerful Tool for the Extraction of Spatial- and Geo-information." *ISPRS Journal of Photogrammetry & Remote Sensing*. Vol. 51, (1996), pg. 63-77.

Bachmann, C., Abdulhai, B., Matthew, Roorda, J., Moshiri, B. (2013) A comparative assessment of multi-sensor data fusion techniques for freeway traffic speed estimation using microsimulation modeling. *Transportation Research Part C: Emerging Technologies* **26**, pp. 33-48, Online publication date: 1-Jan-2013.

Barranquero, J., González, P., Díez, J., José del Coz. J. (2013) On the study of nearest neighbor algorithms for prevalence estimation in binary problems. *Pattern Recognition* **46**:2, pp. 472-482 Online publication date: 1-Feb-2013.

Barzohar, Meir and David B. Cooper. (1995). "New Geometric Stochastic Technology for Finding and Recognizing Roads and Their Features in Aerial Images." A. Gruen, O. Kuebler, P. Agouris (Eds.). *Automatic Extraction of Man-Made Objects from Aerial and Space Images*. Basel, Switzerland: Birkhauser Verlag.

Benediktsson, J.A. and J.R. Sveinsson. (1997). "Feature Extraction for Multisource Data Classification With Artificial Neural Networks." *International Journal of Remote Sensing*. Vol. 18, No. 4, pg. 727-740.

Bennet G., Wolters R. (1996), «A European ecological network», in: *Perspectives on ecological networks,* Arnhem, ECNC publication series on man and nature, voi. I.

Bishop, Ian D., Francisco J. Escobar, Sadasivam Karuppannan, Ian P. Williamson, Paul M. Yates, Ksemsan Suwarnarat, and Haider W. Yaqub. (2000). Spatial Data Infrastructures for Cities. *Cities*. Vol. 17, No. 2, pg. 85-96.

Borowik, T., Pettorelli, N., Sönnichsen, L., Jędrzejewska, B. (2013) Normalized difference vegetation index (NDVI) as a predictor of forage availability for ungulates in forest and field habitats. *European Journal of Wildlife Research* Online publication date: 28-Apr-2013

Breheny, M & R., Rookwood, (2003) Planning the Sustainable City Region, in A. Blowers (ed), Planning for a Sustainable Environment, Earthscan, London, pp 150-189

David Bannister, Kenneth Button, and Peter Nijkamp (Eds.). *Environment, Land Use and Urban Policy.* Cheltenham, United Kingdom: Edward Elgar Publishing Ltd.

Buongiorno, C., Doumaz, F. Spinetti, C., (2010) "Dallo spazio al territorio: un sistema per l'osservazione della Terra in tempo reale" in Geomedia n. 2 pp. 6-10

Burrough, Peter A. (2002). "Possibilities and Constraints of GIS Applications." Pieter van Teeffelen, Leo van Grunsven, and Otta Verkoren, (Eds.). *Possibilities and Constraints of GIS Applications.* Utrecht: Koninklikjk Nederlands Aardrijkskundif Genootschap, pg. 17-25.

Canadian International Development Agency (CIDA). (1998). *An Urbanizing World: Statement on Sustainable Cities.* Hull, Quebec: Minister of Public Works and Government Services Canada Publication.

Canadian International Development Agency (CIDA). (2001). *CIDA's Sustainable Development Strategy 2001-2003.* Hull, Quebec: Minister of Public Works and Government Services Canada Publication.

Canadian International Development Agency (CIDA). (1997). *Our Commitment to Sustainable Development: The Strategy of the Canadian International Development Agency*. Hull, Quebec: Minister of Public Works and Government Services Canada Publication.

Cartwright, T. J. (1992). "Geographic Information Technology as Appropriate Technology for Development." Ian Masser and H. J. Onsrud (Eds.). *Diffusion and Use of Geographic Information Technologies*. Dordecht, The Netherlands: Kluwer Academic Publishers, pg. 261-274.

Castells, Manuel. (1996). *The Rise of the Network Society.* Oxford, United Kingdom: Blackwell Publishers.

C.E.C (Commission of European Community) (1991) *CORINE Biotipes Manual, habitats of the European Community. A model to identify and describe consistently sites of major importance for nature conservation.* EUR 12587/3, Bruxelles, pp. 423.

Cavalli, R. M., Fusilli, L., Pascucci, S., Pignatti, S. Poscolieri, M. (2005): *Mappatura della vegetazione e degli affioramenti rocciosi nel parco del Pollino con dati mivis e hyperion* ASITA 2005

CELADE. (2000). *Knowledge and Data for Population in Development.* Paper Presented at the POPIN Advisory Committee Meeting, ESCAP, Bangkok, June 1995.

Chen, Q., Acey, C., and Lara, J. (2012) Sustainable Futures for Linden Village: A Model for Rehabilitation and Sustainable Development of Urban Neighborhoods. ICSDEC 2012: pp. 163-170.

Ciresan, D. C., Meier, U., Schmidhuber, J.: (2012) Multi-column Deep Neural Networks for Image Classification. IEEE Conf. on Computer Vision and Pattern Recognition CVPR 2012.

Collwell, B.J., (1974), Vegetation Canopy Reflectance, Remote Sensing of Environ., 3: (175-183)

Commissione Europea, (1997), Agenda 2000. Per un'Unione più forte e più ampia, Commissione Europea, COM (97) 2000, Bruxelles, 1997

Couclelis, Helen. (2001). "Requirements for Planning-Relevant GIS: A Spatial Perspective." *Papers in Regional Science: The Journal of the RSAI.* Vol. 70, No. 1, pg. 9-19.

Couclelis, H. (1999). "Spatial Information Technologies and Societal Problems." M. Craglia and H. Onsrud, (Eds.). *Geographic Information Research: Translantic Perspectives*. London: Taylor and Francis, pg. 15-24.

Cowen, David J. and John R. Jensen. (2000). "Extraction and Modeling of Urban Attributes Using Remote Sensing Technology." *People and Pixels: Linking RS and Social Science*. Washington D.C.: National Research Council National Academy Press.

Da Costa, S.M.F. and J.P. Cintra. (2001). "Environmental Analysis of Metropolitan areas." *ISPRS Journal of Photogrammetry & Remote Sensing*. Vol. 54, pg. 41-49.

Darrow K. and R. Pam. (1978). *Appropriate Technology Source Book.* Washington D.C.: Vita.

Debavalya, Nibhon. (1982). "Urbanization and Regional Policies." Victor Sit

Failla A., Porto S.M.C., La Pergola P. (2005): *Schede tecniche di intervento per il recupero di edifici rurali tradizionali nell'area etnea*, in Atti del Convegno AIIA L'ingegneria agraria per lo sviluppo sostenibile dell'area mediterranea, Catania 27-30 Giugno.

Donnay J-P., Barnsley M.J., Longley P.A. (2001), *Remote Sensing and Urban Analysis,* Taylor & Francis, London.

Formica, S. & and Uysal, M. (2000): Market Segmentation of an International Cultural-Historical Event in Italy. *Journal of Travel Research*, 36, 16-24

Fung-shuen and Koichi Mera (Eds.). *Urbanization and National Development: A Comparative Study.* Conference held at the University of Tsukuba, Tsukuba, Japan, 10-12 August 1981, pg. 69-79.

de Man, W.H. Erik. (2000). "Towards Sustainable Geographic Information Technology: Its Institutionalization." Pip Forer, Anthony G.O. Yeh, Jianbang He (Eds.). *Proceedings 9th International Symposium on Spatial Data Handling*. (pg. 6b 22-30). Beijing, People's Republic of China, 10-12 August 2000.

De Roo, G., and D. Miller. (2001). "Introduction – Compact Cities and Sustainable Development." Gert de Roo and Donald Miller (Eds.). *Compact Cities and Sustainable Urban Development.* Hampshire, England: Ashgate Publishing Ltd., pg. 1-16.

Donnay J-P., Barnsley M.J., Longley P.A. (2001), *Remote Sensing and Urban Analysis,* Taylor & Francis, London

Durongdej, Sripen. (2002). *Participatory Planning In Rural Development*. Waterloo, Ontario: PhD Thesis, School of Planning, University of Waterloo.

Dutra, L.V. and R. Huber. (1999). "Feature Extraction and Selection for ERS-1/2 InSAR Classification." *International Journal of Remote Sensing*. Vol. 20, No. 5, pg. 993-1016.

ECNC (European Center for Nature Conservation) (1996), *Perspectives on Ecological Networks,* Arnhem, ECNC publication series on Man and Nature, voi. I.

Edwards, Geoffrey. (2001). "Remote Sensing Image Analysis and Geographic Information Systems: Laying the Groundwork for Total Integration." *Proceedings: The Integration of Remote Sensing and Geographic Information Systems.* Proceedings of a Special Session held at the Baltimore 1991 ACSM-ASPRS Annual Convention.

Elkington, John and Jonathan Shopley. (2000). *The Shrinking Planet: U.S. Information Technology and Sustainable Development*. Pennsylvania, U.S.A.: World Resources Institute.

Entwisle, Barbara, Stephen J. Walsh, Ronald R. Rindfuss, and Aphichat Chamratrithirong (2004). "Land-Use/Land-Cover and Population Dynamics" *People and Pixels: Linking RS and Social Science*. Washington D.C.: National Research Council National Academy Press.

Fiset, Robert, Francois Cavayas, Marie-Catherine Mouchot, Basel Solaiman, and Robert

Desjardins. (1998). "Map-image Matching Using a Multi-layer Perception: the Case of the Road Network." *ISPRS Journal of Photogrammetry & Remote Sensing*. Vol. 53, (1998), pg. 76-84.

Di Paola G. (2004), *Presente e futuro del Parco nel suo Sistema Informativo*, in "Bollettino trimestrale dell'ente Parco dell'Etna", n. 11.

Di Paola (2006). "Il piano territoriale del Parco dell'Etna e la pianificazione strategica come strumento per lo sviluppo sostenibile" in *La storia del primo Parco istituito in Sicilia dalle pagine del suo Bollettino,* (2006). pp. 204-207.

Fasel Q. and Khan K. A, (2012): Investigations of Cellular Automata Linear Rules for Edge Detection, I. J. Computer Network and Information Security, Vol. 3, pp. 47-53

Fisher, R., S. Perkins, A. Walker, and E. Wolfart. (2000). *Hypermedia Image Processing References (HIPR2).* http://www.dai.ed.ac.uk/HIPR2/. Accessed October, 2001.

Foody, Giles M. (2005). "Land Cover Classification by an Artificial Neural Network with Ancillary Information." *International Journal of Geographical Information Systems.* Vol. 9, No. 5, pg. 527-542.

Foody, Giles M. (2004). "The Significance of Border Training Patterns in Classification by a Feedforward Neural Network Using Back Propagation Learning." *International Journal of Remote Sensing.* Vol. 20, No. 18, pg. 3549-3562.

Frallicciardi A.M., Delle Donne B., Graniero M., Palmentieri S. (2001), II Parco Nazionale del Vesuvio e il Parco Regionale dei Campi Flegrei: due nodi nelle politiche di sviluppo sostenibile regionale, in: Menegatti B., Tinacci Mossello M. (a cura di) *Sviluppo sostenibile a scala regionale,* Bologna, Patron,, pp.476-506.

Freund, John E. and Gary A. Simon. 1992. *Modern Elementary Statistics.* New Jersey: Prentice-Hall Inc.

Gahegan, Mark and Manfred Ehlers. (2000). A Framework for the Modeling of Uncertainty Between Remote Sensing and Geographic Information Systems. *ISPRS Journal of Photogrammetry & Remote Sensing.* Vol. 55 (2000), pg. 176-188.

Ghose, Rina (2011) Politics and power in the use of GIS for community decision making. pp. 423-438 in Timothy Nyerges, Helen Couclelis, Robert McMaster (eds)

Handbook of GIS and Society Research.

Ghose, R, Global Development Research Center (GDRC). (2001). Introduction: Urban Environmental Management." *The WWW Virtual Library, Urban Environmental Management*. http://www.gdrc.org/uem/index.html. Accessed October, 2001.

Gomarasca M.A., (2000) Introduzione a Telerilevamento e GIS per la gestione delle risorse agricole e ambientali, Ed. AIT, (1 Ed. 1997)

Gomarasca M.A., (2004) Elementi di Geomatica, ed. AIT

Gruen, Armin and Haihong Li. (2002). "Road Extraction From Aerial and Satellite Image by Dynamic Programming." *ISPRS Journal of Photogrammetry and Remote Sensing.* Vol. 50, No. 4, pg. 11-20.

Gruen, Armin and Haihong Li. (2002). "Semi-Automatic Linear Feature Extraction by Dynamic Programming and LSB-Snakes." *Photogrammetric Engineering and Remote Sensing*. Vol. 63, No. 8, pg. 985-995.

Gruen, Armin, Peggy Agouris, and Haihong Li. (2003). "Linear Feature Extraction with Dynamic Programming and Globally Enforced Least Squares Matching." A. Gruen, O. Kuebler, P. Agouris (Eds.). *Automatic Extraction of Man-Made Objects from Aerial and Space Images*. Basel, Switzerland: Birkhauser Verlag.

Gruen, Armin, Peggy Agouris, Dirk Stallmann, and Haihong Li. (2004). "Algorithms for Automated Extraction of Man-Made Objects from Raster Image Data in a GIS." Jurg Nievergelt, Thomas Roos, Hans-Jorg Schek, and Peter Widmayer (Eds.). From IGIS '94: Geographic Information Systems, proceeding from a workshop on Advanced Research in Geographic Information Systems, Monte Verita, Ascona, Switzerland,

February/March 1994. *Lecture Notes in Computer Science*. Berlin, Germany: Springer-Verlag, pg. 123-132.

Guindon, B. (2000). "A Framework for the Development and Assessment of Ojbect Recognition Modules from High-Resolution Satellite Images". *Canadian Journal of Remote Sensing*. Vol. 26, No. 4. pg. 334-348.

Haala, Norbert and Claus Brenner. (1999). "Extraction of Buildings and Trees in Urban Environments." *ISPRS Journal of Photogrammetry & Remote Sensing*. Vol. 54 (1999), pg. 130-137.

Haklay, Mordechai (Muki); Francis, Louise; Hanif Rahemtulla; and Claire Ellul (2010) Mapping for change. GIS @Development 14 (2) 34-37. (Neogeography & Participatory GIS: Power to people) February, http://www.gisdevelopment.net/magazine/global/

Hall, G. Brent. (2003). *GIS Use in Rural Planning and Management*. Keynote Address at Second Seminar on GIS (GISDECO): The Practice of GIS Applications: Problems and Challenges. Soustenberg, The Netherlands, June 1993.

Hall, G.B. (1999). "Guest Editorial - GIS Education and Infrastructure Challenges and Problems." *Transactions in Geographic Information Systems.* Vol. 3, No. 4, pg. 311-318.

Hall, G. Brent and Arthur Conning. (1992). "Integrating Demographic and Socio-economic Data with PC-Based GIS" Pieter van Teeffelen, Leo van Grunsven, and Otta Verkoren, (Eds.). *Possibilities and Constraints of GIS Applications s.* Utrecht: Koninklikjk Nederlands Aardrijkskundif Genootschap, pg. 87-102.

Hall, G.B., F. Wang, and Subaryono. (1992). "Comparison of Boolean and Fuzzy Classification Methods in Land Suitability Analysis by Using Geographical Information Systems." *Environment and Planning A*. Vol. 24, pg. 497-516.

Hall, G.B., N.W. Malcolm, and P.M. Piwowar. (2001). "Integration of Remote Sensing and GIS" *Transactions in GIS.* Vol. 5, No. 3, pg. 235-254.

Hanson, J. and U. Narula. (1990). *New Communication Technologies*. Hillside, New Jersey: LEA. http://www.sister.it/itaCorine/Corine/legendadettagliata.htm

A. Hauschild, O. Montenbruck, P. Steigenberger, W. Cao, R.B. Langley, L. Urquhart, and M. Santos (2010): Evaluating the Performance of GPS/GIOVE Integrated Precise Point Positioning, GPS Wold Magazine, 12 Jul 2010, http://www.questexenews.com/t.do?id=5231470:16878914

Hoctor T. S.; Carr M. H. (2) ; Zwick P. D (2000) Identifying a linked reserve system using a regional landscape approach: The Florida Ecological network *Journal of Conservation Biology* vol. 14, n°4, pp. 984-1000

Huan, Vu Duc. (1995). *Application of Remote Sensing and Geo-information System for Evaluating Urban Development*. Thesis No. NR-95-3. Space Technology Applications and Research Program, Asian Institute of Technology, Bangkok, Thailand.

Hunziker, M., Kienast, F., 1999. Potential impact of changing agricultural activities on scenic beauty – a prototypical technique for automated rapid assessment. Lands. Ecol. 14, 161–176.

Huxhold, William E., and Allan G. Levinsohn. (2000). *Managing Geographic Information System Projects.* New York, New York: Oxford University Press, Inc.

Jacobelli P., Fiduccia A., Allegrini A., Colletta F., Novelli L. (2004) *Il telerilevamento iperspettrale Mivis a supporto della costruzione di processi innovativi di planning urbano-territoriale*, Atti della VIII Conferenza nazionale Asita, vol. II, pp. 1285 -1290.

Jankowski, Piotr; and Timothy L. Nyerges (2001) GIS-supported collaborative decisionmaking: results of an experiment. Annals of Assoc. of American Geographers 91, pp 48-70.

International Council for Local Environmental Initiatives (ICLEI) and the International

Development Agency (IDRC). (1996). *The Local Agenda 21 Planning Guide: An Introduction to Sustainable Development Planning*. Toronto, Canada: The International Council for Local Environmental Initiatives (ICLEI).

Intesa Stato – Regioni – Enti locali per la realizzazione dei sistemi informative geografici (Versione 16I del 29 Aprile 2001) http://www.intesagis.it/specifiche_tecniche.asp

ISTAT (2001), 14 Censimento della popolazione e delle abitazioni. Roma

Jedynak, Bruno and Jean-Philippe Roze. (1995). "Tracking Roads in Satellite Images by Playing Twenty Questions." A. Gruen, O. Kuebler, P. Agouris (Eds.). *Automatic Extraction of Man-Made Objects from Aerial and Space Images*. Basel, Switzerland: Birkhauser Verlag.

Jensen, John R. (1996). *Introductory Digital Image Processing: A Remote Sensing Perspective*. Upper Saddle River, New Jersey: Prentice Hall, Inc. Jiang, Bin. (2000). "Agent-Based Approach to Modelling Environmental and Urban Systems Within GIS." Pip Forer, Anthony G.O. Yeh, Jianbang He (Eds.). *Proceedings 9th International Symposium on Spatial Data Handling*. (pg. 3b 17-29). Beijing, People's Republic of China, 10-12 August 2000.

Johnson, Craig. (2000). "Common Property, Political Economy, and Local Resource Management." *South East Asia Research*. Vol. 8, No. 1, pg. 31-53.

Jongman R.H.G., (1997) *Le reti ecologiche: a quale scopo? Elementi naturali indispensabili.* E anche Negrini G., *La rete ecologica Europea*, in: Peano A., Parchi naturali in Europa, Urbanistica Dossier, 7, INU, Roma

Kahila, Maarit; and Marketta Kyttä (2010) SoftGIS as a bridge-builder in collaborative urban planning. Journal of Planning Support Systems - Best Practice and New Methods, pp. 389–411.

Kalluri, S.N.V., J. Jaja, D.A. Bader, Z. Zhang, J.R.G. Townshend, and H. Fallah-Adl. (2000). "High Performance Computing Algorithms for Land Cover Dynamics Using Remote Sensing Data." *International Journal of Remote Sensing*. Vol. 21, No. 6 & 7, pp. 1513-1536.

Kamaludin, H., Lihan, T., Ali Rahman, Z., Mustapha, M. A., Idris, W. M. R., and Rahim, S. A. (2013): Integration of remote sensing, RUSLE and GIS to model potential soil loss and sediment yield (SY), *Hydrol. Earth Syst. Sci. Discuss.*, 10, pp 4567-4596

Karatas, A. and El-Rayes, K. (2012) Evaluating and Quantifying Sustainable Development in Urban Neighborhoods. ICSDEC 2012: pp. 214-221.

Karimi, Hassan A., Xiaolong Dai, Siamak Khorram, Aemal J. Khattak, and Joseph E. Hummer. (2003). "Techniques for Automated Extraction of Roadway Inventory Features from High-Resolution Satellite Imagery." *Geocarto International*. Vol. 14, No. 2, pg. 5-16.

Kavzoglu, T., Colkesen, I.. (2013) An assessment of the effectiveness of a rotation forest ensemble for land-use and land-cover mapping. *International Journal of Remote Sensing* **34**:12, pp 4224-4241. Online publication date: 20-Jun-2013.

Kim, Taejung and Jan-Peter Muller. (2001). "Building Extraction and Verification from Spaceborne and Aerial Imagery using Image Understanding Fusion Techniques." A. Gruen, O. Kuebler, P. Agouris (Eds.). *Automatic Extraction of Man-Made Objects from Aerial and Space Images*. Basel, Switzerland: Birkhauser Verlag.

Kubota, Satoshi; Kazuya Soga, Yuki Sasaki and Akihiro Abe (2012): Web GIS-based regional social networking service as participatory GIS. Recent Progress in Data Engineering and Internet Technology. Lecture Notes in Electrical Engineering 157, pp 313-321

Laquian, Aprodicio A. (2000). *The Planning and Goverence of Mega-urban Regions: What Can We Learn?* Vancouver, Canada: Centre for Human Settlements, University of British Columbia, Vancouver, B.C. From the ACSP Conference, Atlanta, November 2,

2000. http://www.chs.ubc.ca/publications/Mur/Mega-urban_Regions.html (Accessed March 2001).

Li, Xia and Anthony Gar-On Yeh. (2000). "Modelling Sustainable Urban Development by the Integration of Constrained Cellular Automata and GIS." International Journal of Geographical Information Science. Vol. 14, No. 2, pg. 131-152.

Lillesand, T. and R. W. Kiefer. (1994). Remote Sensing and Image Interpretation. New York: John Wiley & Sons, Inc.

Lo Tauro, A. & E.F. Brown: *Growth and development process in urban system La morfologia nello studio dei sisrtemi urbani* in AISRe Pre-Print Book, Siena "Le Scienze Regionali" 30-31Oct -1 Nov. 1995.pp 334-361

Lo Tauro, A: *Cambiamento di tendenza nel concetto di Spazi Aperti*, Corso di Urbanistica I A.A 1994-95, Univerita' di Palermo, Facolta' di Architettura. PP 1-45

Lo Tauro, A (1995) Morphological implications in urban analysis. Accepted for publication in Urban Morphology Newsletter, The Urban Morphology Research group, University of Birmingham, School of Geography pp 11

Lo Tauro, A. (1998) *La tecnologia della pianificazione strategica* e *Politiche di sviluppo regionali* in AIRSe news n. 16 pg.10 e 17

Lo Tauro A. (2001): Beni culturali, Attività produttive, valenze ambientali, Master Thesis in "Storia ed analisi del territorio", Scuola Superiore, Catania University

Lo Tauro, A (2002) *Sustainable development and markets in the southern Italy*, International Symposium on Environmental Issues and Waste Management, SWEMP 2002 (Caglairi, 7-10 Ottobre 2002) CD-ROM pp. 79-85

Lo Tauro A. (2003): Technologies for Cultural Haritage: cultural electronic portals and webGIS, Master Thesis in "Economia del recupero e della valorizzazione dei beni culturali", Scuola Superiore, Catania University

Lo Tauro, A (2003) *Sustainable strategies in Sicilian towns*, *THE SUSTAINABLE CITY* Second International Conference on Urban Regeneration and Sustainability, CONFERENCE and ROYAL FORMAT BOOKS Witt Press Southamton, Boston

Lo Tauro A. e Candura A.R., *Caratterizzazione dei parchi tramite il clc 90 2000: il parco dell'Etna*, in *Atti della 8a Conferenza italiana UtentiESRI* (Roma, 20-21 aprile 2005), Roma, ESRI, 2005, CD-ROM (esriitalia.it/conferenza2005/).

Lo Tauro, A, Di Paola, G. Spina, S.E. (2005) *Preservation and Valorisation of Cultural heritage and information systems, an investigation into a web-GIS* in MULTIMEDIAPLAN.AT, CORP&Geomultimedia05 (feb.22-25 2005, www.corp.at) pp. 311-318

Lo Tauro, A. (2005) – Tecnologie GIS per la tutela e valorizzazione dei beni culturali", *Sessione 2.1 GIS/applicazioni nella pianificazione e gestione territoriale"*, in *Atti della 9[9] Conferenza ASITA*, pp. 1375 – 1380

Lo Tauro A, Allegrini A., Di Paola, G. Spina, S.E (2005a) *Pattern recognition e remote sensing: il territorio del Parco dell'Etna* in AreaAvasta, Anno 6, Numero 10/11 (rivista semestrale della Provincia di Salerno - pp 142-148

Lo Tauro, A. e Costa N. (2005c) "Trasformazioni territoriali e VIA. Un modello di analisi su piattaforma GIS", *Sessione 2.1 GIS/applicazioni nella pianificazione e gestione territoriale"*, in *Atti della 9[9] Conferenza ASITA*, pp. 1381-1386

Lo Tauro, A. (2006) "Remote sensing and GIS for the implementation of thematic cartography in protected areas*" in Econgeo2006 - 5th European Congress on Regional Geoscientific Cartography and Information Systems*, Session 10, Maps, Information Systems and New Tools, June 13th to 16th 2006. CD ROM

Lo Tauro, A. (2007). The integration of Remote Sensing and GIS to facilitate Environmental and Cultural Resources Management. Ph.D Thesis, Faculty of Informatics and Mathematics, University of Trieste, pp. 1-200.

Lo Tauro, A. (2010) Geomatics: Research of Novel Applications in Heritage Management, in Vector1 Media: Promoting Spatial Analysis for Sustainable Development, (http://www.vector1media.com/articles/features/11520-geomatics-research-of-novel-applications-in-heritage-management)

Lo Tauro, A. (2010 a) Ricognizione aerea, fotografia digitale ed elaborazione GIS based per la cartografia tematica del paesaggio e del patrimonio culturale in GEOmedia http://www.rivistageomedia.it/Approfondimenti/ricognizione-aerea-fotografia-digitale-ed-elaborazione-gis-based-per-la-cartografia-tematica-del-paesaggio-e-del-patrimonio-culturale.html

Lo Tauro, A. (2012) Geospatial Analysis and Data Integration for Cultural Resources Evaluation: A collection of articles on analytical geomatics and their applications. BAR International Series 2313 , 2011 pp. 90 (http://www.barpublishing.com)

Low, Nicholas, Brendan Gleeson, Ingemar Elander and Rolf Lidskog. (2000). Nicholas Low, Brendan Gleeson, Ingemar Elander, and Rolf Lidskog (Eds.). *Consuming Cities: The Urban Environment in the Global Economy after the Rio Declaration.* London, England: Routledge, pg. 281-307.

Mahesh Pal, Giles M Foody. (2010) Feature Selection for Classification of Hyperspectral Data by SVM. *IEEE Transactions on Geoscience and Remote Sensing* 48:5, 2297-2307 Online publication date: 1-May-2010.

Makarovic, B. (2003). "Image-based Digital Mapping." *ISPRS Journal of Photogrammetry and Remote Sensing*. Vol. 50, No. 2, pg. 21-29.

Malcolm, N.W., J.M. Piwowar, G.B. Hall, C. Cotlier, and A. Ravenna. (2001). "An Integration of RADARSAT and Landsat Imagery" *Canadian Journal of Remote Sensing*. Vol. 27, No. 6,

Manzoni, G., Palermo, C., 2003. Il Catasto delle Strade nel progetto pilota della Provincia di Teramo, *Ondaverde,* 83, pp. XI-XII.

Manzoni, G. 2006, Telegeomatica applicata al rischio vulcanico. In *Atti del Convegno "Il piano Etna, rischio vulcanico versante sud", Nicolosi, 12 giugno 2006*

Mark, David M., Nicholas Chrisman, Andrew U. Frank, Patrick H. McHaffie, and John Pickles. (1997). *The GIS History Project.*
http://www.geog.buffalo.edu/ncgia/gishist/bar_harbor.html (Assessed March 2001).

Martinez-Casasnovas, Jose A. (2000). "A Cartographic and Database Approach for Land Cover/Use Mapping and Generalization from Remotely Sensed Data." *International Journal of Remote Sensing.* Vol. 21, No. 9, pg. 1825-1842.

Masser, I. and H. Campbell. (1989). *Conditions for the Effective Utilization of Computers in Urban Planning.* United Kingdom: University of Sheffield, TRP 88, Department of Town and Country Planning.

McGill, Ron. (2001). "McGill Management Checklist." Cities. Vol. 18, No. 5, pg. 347-354.

Menou, Michel J. (1993). *Measuring the Impact of Information on Development.* Ottawa, Canada: International Development Research Council (IDRC).

Mahesh Pal, Aaron E. Maxwell, Timothy A. Warner. (2013) Kernel-based extreme learning machine for remote-sensing image classification. *Remote Sensing Letters* 4:9, pp 853-862

Micelli, S. & Di Maria, E. (2000): *Città e cultura nell'economia delle reti*, Bologna. Il Mulino

Ming-Hsiang Tsou and Ken Yanow (2010) Enhancing General Education with Geographic Information Science and Spatial Literacy. *URISA Journal* • Vol. 22, No. 2 • 2010

Mitchell, A., (1999) *GIS Analysis*, Redlands, Esri Press Molenaar, M. and L.L.F. Janssen. (2003). "Integrated Processing of Remotely Sensed and Geographic Data for Land Inventory Purposes." *Proceedings: The Integration of Remote Sensing and Geographic*

Information Systems. Proceedings of a Special Session held at the Baltimore 2003 ACSM-ASPRS Annual Convention.

Morain, S. (1998). "A Brief History of Remote Sensing Applications, with Emphasis on Landsat." David Liverman, Emilio F. Moran, Ronald R. Rindfuss, and Paul C. Stern (Eds.). *People and Pixels: Linking Remote Sensing and Social Science*. Washington D.C.: National Academy Press.

Muhar, A. (1999) Three dimensional modelling and visualisation of vegetation for lanscape simulation, special issue of Landscape and urban planning, *Our Visual Landscape Conference*, Ascona 1999

Musumeci, M. (2000) *Innovazione Tecnologica e Beni Culturali. Uno studio sulla situazione della Sicilia*, Cagliari, CRENoS

Musson C., Palmer R., Stefano C. (2005): *In volo nel Passato, aerofotografia e cartografia archeologica*, All'Insegna del Giglio, Biblioteca del Dipartimento di Archeologia e Storia dell arti – sezione archeologica, Università di Siena

Nedovic-Budic, Zorica. (1997). "GIS Technology and Organizational Context: Interaction and Adaption." M. Craglia and H. Couclelis (Eds.). *Geographic Information Research*. London, England: Taylor and Francis.

Nghi, D. Q. and H. D. Kammeier. (2001). "Balancing Data Integration Needs in Urban Planning." *Cities*. Vol. 18, No. 2, pg. 61-75.

Ngwainmbi, Emmanuel K. (1999). *Exporting Communication Technology to Developing Countries*. Lanham, Maryland: University Press of America.

Nocentini S. (2004), *Parchi e riserve di varia tipologia e alter aree protette*, in *Italia. Atlante dei tipi geografici*, Firenze, IGM, pp. 391-394.

Yıldız, O. T. (2011) Mapping classifiers and datasets. *Expert Systems with Applications* 38:4, pp 3697-3702, Online publication date: 1-Apr-2011.

Ogawa, Naohiro. (1985). "Urbanization and Internal Migration in Selected ASEAN Countries: Trends and Prospects." Philip M. Hauser, Daniel B. Suits, and Naohiro Ogawa (Eds.). *Urbanization and Migration in Asean Development*. Tokyo, Japan: National Institute for Research Advancement, pg. 83-108.

Nuryanti, W. (2004): Heritage and Postmodern tourism. *Annals of Tourism Research* 23(2), 249-260

Onchan, Tongroj. (2002). "Urbanization and Life Environment Protection" *International Symposium on APEC for Sustainable Development*. Tokyo, Japan

O'Sullivan, David, Alastair Morrison, and John Shearer. (2000). "Using Desktop GIS for the Investigation of Accessibility by Public Transport: an isochrone approach." *International Journal of Geographical Information Science*. Vol. 14, No. 1, pg. 85-104.

Ouma-Onyango, Richard. (1997). *Information Resources and Technology Transfer Management in Developing Countries*. London, England: Routledge.

Pal, S.K., A. Ghosh, and B. Uma Shankar. (2000). "Segmentation of Remotely Sensed Images with Fuzzy Thresholding, and Quantitative Evaluation." *International Journal of Remote Sensing*. Vol. 21, No. 11, pg. 2269-2300.

Pereira, L.M. Gomes and L.L.F. Janssen. (1999). "Suitability of Laser Data for DTM Generation: a Case Study in the Context of Road Planning and Design." *ISPRS Journal of Photogrammetry & Remote Sensing*. Vol. 54 (1999), pg. 244-253.

Pesaresi C. (2003), *Le recenti eruzioni dell'Etna: danni materiali, interventi adottati, fenomeni anomali*, "Semestrale di Studi e Ricerche di Geografia", Roma, Tip. Abilgraph, 1, pp. 149-195.

Petit, C.C. and E.F. Lambin. (2001). "Integration of Multi-source Remote Sensing Data for Land Cover Change Detection." *International Journal of Geographical Information Science*. Vol. 15, No. 8., pg. 785-803.

Pilati, A. (2002) Nuove tecnologie: Opportunità e Mercati, *Economia della Cultura*, (I), 35-37

Portnov, Boris A. and David Pearlmutter. (1999). "Sustainable Urban Growth in Peripheral Areas." *Progress in Planning*. Vol. 52 (1999), pg. 239-308.

Powell, R., Bailey, R., Wade, A., and Tunstall, J. (2012) Reaching Out to Local Communities in a Discussion of Sustainable Development. ICSDEC 2012: pp. 180-187.

Pratibha Sharma, Manoj Diwakar and Niranjan Lal. (2013): Edge Detection using Moore Neighborhood. *International Journal of Computer Applications* 61(3):26-30, January 2013. Published by Foundation of Computer Science, New York, USA

Protezione Civile, Comune di Nicolosi (2006) *Piano Enta: rischio vulcanico versante sud Nicolosi*, edizione giugno 2006, Nicolosi, CD ROM

Provincia Regionale di Catania (2003), *L'area metropolitana di Catania*, Tipolitografia Etna, Catania Provincia Regionale di Catania (2003), *Schema di massima del piano Territoriale*, Catania (www.provincia.ct.it)

Puglisi, A. (2006): "Gestione ambientale: primi passi per lo sviluppo del sistema" in *La storia del primo Parco istituito in Sicilia dalle pagine del suo Bollettino,* (2006). pg. 208-209.

Punpuing, Sureeporn and Helen Ross. (2001). "Commuting." *Cities*. Vol. 18, No. 1, pg. 43-50.

Ragia, Lemonia and Stephan Winter. (2000). "Contributions to a Quality Description of Areal Objects in Spatial Data Sets." *ISPRS Journal of Photogrammetry & Remote Sensing*. Vol. 55 (2000), pg. 201-213.

Ramasubramanian, Laxmi. (2003). "GIS Implementation: Learning from Organizational Theory and Reflective Practice." *Transactions in Geographic Information Systems*. Vol. 3, No. 4, pg. 359-380.

Regional Municipality of Waterloo. (1998). *Regional Official Policies Plan (ROPP)*. Waterloo, Ontario: Region of Waterloo, December 1998 Consolidation as approved by the Minister of Municipal Affairs and Housing, December 19, 1995.

Regolamento CEE n. 2081/1992, G.U.C.E. n. L. 208 del 24/07/1992

Ricotta, C. and G.C. Avena. (1999). "The Influence of Fuzzy Set Theory on the Areal Extent of Thematic Map Classes." *International Journal of Remote Sensing*. Vol. 20, No. 1, pg. 201-205.

Richards, G. (2003): Production and Consumption of European Cultural Tourism. *Annals of Tourism Research*, 22(2) 261-283

Richards, John A. and Xiuping Jia. (1999). *Remote Sensing Digital Image Analysis: An Introduction. Third, Revised and Enlarged Edition*. Berlin, Germany: Springer.

Russ, John C. (1995). *The Image Processing Handbook. 2nd Edition*. Boca Raton, Florida: CRC Press Inc.

Saraf, A.K. (1999). "IRS-1C-LISS-III and PAN Data Fusion: An Approach to Improve Remote Sensing Based Mapping Techniques." *International Journal of Remote Sensing*. Vol. 20, No. 10, pp. 1929-1934.

Satterthwaite, David. (2001). "Sustainable Cities or Cities that Contribute to Sustainable Development." David Bannister, Kenneth Button, and Peter Nijkamp (Eds.). *Environment, Land Use and Urban Policy*. Cheltenham, United Kingdom: Edward Elgar Publishing Ltd.

Shaban, M.A. and O. Dikshit. (2002). "Evaluation of the Merging of SPOT Multispectral and Panchromatic Data for Classification of an Urban Environment." *International Journal of Remote Sensing*. Vol. 23, No. 2, pg. 249-262.

Shafer, C.S., Scott, D., Baker, J. & Winemiller, K., (2013), Recreation and Amenity Values of Urban Stream Corridors: Implications for Green Infrastructure, *Journal of Urban Design*

Sheppard, Eric, Helen Couclelis, Stephen Graham, J.W. Harrington, and Harlan Onsrud. (1999). "Geographies of the Information Society." *International Journal of Geographical Information Science.* Vol. 13, No. 8, pg. 797-823.

Shettigara, Vittala K., Siegfried G. Kempinger, and Robert Aitchison. (2000). "Semi-Automatic Detection and Extraction of Man-Made Objects in Multispectral Aerial and Satellite Images." A. Gruen, O. Kuebler, P. Agouris (Eds.). *Automatic Extraction of Man-Made Objects from Aerial and Space Images.* Basel, Switzerland: Birkhauser Verlag.

Shirowzhan, S. and Lim, S. (2012) Extraction of Polygon Footprints from Lidar Data in an Urban Environment. ICSDEC 2012: pp. 246-251

Shirowzhan, S. and Lim, S. (2012) Development of Urban Sustainability Index Using 3-D Spatial Metrics. ICSDEC 2012: pp. 238-245.

Silberberg, T. (2003): Cultural tourism and business opportunities for museums and heritage sities. *Tourism Management*, 16(5), 361-365

Sinclair, Stuart. (1982). *The Third World Economic Handbook.* London, England: Euromonitor Publications Ltd.

Sit, Victor F.S. (1982). "Urbanization and Development: An Introduction." Victor Sit Fung-shuen and Koichi Mera (Eds.). *Urbanization and National Development in Asia: A Comparative Study.* Conference held at the University of Tsukuba, Tsukuba, Japan, 10-12 August 1981, pg. 1-6.

Sowmya, Arcot and John Trinder. (2000). "Modeling and Representation Issues in Automated Feature Extraction from Aerial and Satellite Images." *ISPRS Journal of Photogrammetry & Remote Sensing.* Vol. 55 (2000), pg. 34-47.

Srinivas, Hari. (1997). "Information Systems in Urban Environmental Management: Roles for the Internet." Paper presented at the Second International Symposium on Urban Planning and Environment on Strategies and Methods for Improving Environmental Quality in Compact Cities. 11-14 March 1997. Groningen, The Netherlands.

Stavrakoudis, D.G., Theocharis, J.B., Zalidis. G.C. (2010): A multistage genetic fuzzy classifier for land cover classification from satellite imagery. *Soft Computing* Online publication date: 17-Nov-2010.

Steger, Carsten, Clemens Glock, Wolfgang Eckstein, Helmut Mayer, and Bernd Radig. (1995). "Model-Based Road Extraction from Images." A. Gruen, O. Kuebler, P.

Agouris (Eds.). *Automatic Extraction of Man-Made Objects from Aerial and Space Images.* Basel, Switzerland: Birkhauser Verlag.

Steinle. E., Voegtle, T. 2001 *Automated extraction and reconstruction of buildings in laser scanning data for disaster management* Third International Workshop on Automatic Extraction of Man-Made Objects from Aerial and Space Images, June 10-15, Ascona, Switzerland

Stow, Douglas, Sally Westmoreland, David McKinsey, Robert Parrott, Sue Carnevale and Doretta Collins. (2001). "Processing Flows for Correcting and Updating Vectorcoded GIS Layers using Remotely Sensed Data." *Proceedings: The Integration of Remote Sensing and Geographic Information Systems.* Proceedings of a Special Session held at the Baltimore 2001 ACSM-ASPRS Annual Convention.

Sui, Daniel Z. (1998). "GIS-based Urban Modelling: Practices, Problems, and Prospects" *International Journal of Geographical Information Science.* Vol. 12, No. 7, pg. 651-671.

Surace, L. (1998), La georeferenziazione delle informazioni territoriali. *Bollettino di Geodesia e Scienze Affini dell'Istituto Geografico Militare*, 58, pp. 181-234.

Susca S. (2010).: **"**GNSS-independent Navigation Solution Using Integrated LiDAR Data,**"** In Proc. of *IEEE/ ION Technical Meeting 2010*, San Diego, CA, USA, January 2010.

Tanaka, Sotaro and Toshiro Sugimura. (2001). "A New Frontier of Remote Sensing from IKONOS Images." ." *International Journal of Remote Sensing.* Vol. 22, No. 1, pg. 1-5.

D.W. Rhind (Eds.). *Geographical Information Systems: Principals and Applications Volume 2.* London, England: Longman, pg. 71-84.

Thomson, Curtis N. and Perry Hardin. (2000). "Remote Sensing /GIS Integration To Identify Potential Low-income Housing Sites." *Cities.* Vol. 17, No. 2, pg. 97-109.

Tirasawat, Penporn. (1985) "Migration in Thailand: Past and Present." Philip M. Hauser, Daniel B. Suits, and Naohiro Ogawa (Eds.). *Urbanization and Migration in Asean Development.* Tokyo, Japan: National Institute for Research Advancement, pg. 407-425.

Trinder, Dr. John. (1995). "Workshop on Automated Extraction of Man-made Objects from Aerial and Space Images." *ISPRS Journal of Photogrammetry & Remote Sensing.* Vol. 50, No. 5, pg. 33-34.

Tripathi, N.K., K.V.G.K. Gokhale, and M.U. Siddiqui. (2000). "Directional Morphological Image Transforms for Lineament Extraction from Remotely Sensed Images." *International Journal of Remote Sensing.* Vol. 21, No. 17, pg. 3281-3292.

Tjatera, I Wayan. (1994). *A Model for Integrating Sub-Provincial Traditional and Official Regional Planning Institutions To Achieve Sustainable Development.* Waterloo, Ontario: PhD Thesis, Regional Planning and Resource Development, University of Waterloo.

Tucker, C. J., (1980). Remote Sensing of leaf water content in the near infrared, *Remote Sensing Environment.* 10:23-32.

Ulbricht, K.A., H.S. Teotia, W.D. Heckendorff, and D.L. Civco. (2001). "International Scientific Cooperation: Application of Remote Sensing for Development Purposes." *ISPRS Journal of Photogrammetry & Remote Sensing.* Vol. 50, No. 5, pg. 31-32.

Ulbricht, K.A. and W.D. Heckendorff. (2002). "Satellite Images for Recognition of Landscape and Landuse Changes." *ISPRS Journal of Photogrammetry & Remote Sensing.* Vol. 53 (2002), pg. 235-243.

United Nations (1996). *Decision support systems for sustainable development: experience and potential* United Nations University, Institute for Software Technology, Macau.

United Nations. (1992). "Information for Decision-Making." *Agenda 21, Chapter 40.* United Nations Conference on Environment & Development, as adopted by the

Plenary in Rio de Janeiro, June 1992.

United Nations Commission on Science and Technology for Develoment (UNCSTD). (2002). *Knowledge Societies: Information Technology for Sustainable Development.* Robin

Mansell and Uta When (Eds.). New York: Oxford University Press Inc. University Consortium for Geographic Information Science (UCGIS). (1998). *GIS and Society.* http://www.geog.umn.edu/umucgis/research_priorities/GISOCIET.html

University Consortium for Geographic Information Science (UCGIS). (1998). *GIS and*

Society II. http://www.geog.umn.edu/umucgis/research_priorities/GISandsocII.pdf

Urban Environmental Management Program (UEM). (2001). "Program Objectives." *Urban Environmental Management Field of Study.* http://www.uem.ait.ac.th/.

Vanschoren, J., Blockeel, H., Pfahringer, B., Holmes, G.. (2012) Experiment databases. *Machine Learning*, Online publication date: 5-Jan-2012.

Vidal, A., Pinglo, F., Durand, H., Devauxros, C., & Maillet, (1994). A. Evaluation of a Temporal Fire Risk Index in Mediterranean Forests From Noaa Thermal Ir.. *Remote Sensing of Environment.* 49(3):296-303. 10 © 2002 Millpress, Rotterdam, ISBN 90-77017-72-0

Viegas, D. X., Bovio, G., Ferreira, A., Nosenzo, A., and Sol, B. (2000). Comparative Study of Various Methods of Fire Danger Evaluation in Southern Europe. *International Journal of Wildland Fire* 9:235-246.

Villa, B. (2005). Tecniche Innovative di rilievo e rappresentazione dei Beni Culturali, Corso di Aggiornamento SIFET 2005.

Viles RL, Rosier DJ, (2001) "How to use roads in the creation of greenways: case studies in three New Zealand landscape", Landscape and Urban Planning, Vol. 55, No.1, pp.15-27, June 15, 2001.

Vosselman, G., Suveg, 1., 2001 *Map based building reconstruction from laser data and images* Third International Workshop on Automatic Extraction of Man-Made Objects from Aerial and Space Images, June 10-15, Ascona, Switzerland

Vosselman, G., Dijkman, S., 2001 '3D building model reconstruction from point clouds and ground plans' IAPRS, Voi. 34, pp. 37-43

Vosselman, George and Jurrien de Knecht. (2002). "Road Tracing by Profile Matching and Kalman Filtering." A. Gruen, O. Kuebler, P. Agouris (Eds.). *Automatic Extraction of Man-Made Objects from Aerial and Space Images*. Basel, Switzerland: Birkhauser Verlag.

Weber T., Wolf J. (2004) Landscape Ecological Assessment of the Chesapeake Bay Watershed Journal of Environmental Monitoring and Assessment Volume 94, Numbers 1-3 / June, 2004 pp 39-53

Wehr, Aloysius and Uwe Lohr. (1999). "Airborne Laser Scanning – An Introduction and Overview." *ISPRS Journal of Photogrammetry & Remote Sensing.* Vol. 54 (1999), pg. 68-82.

Weimin, Zhang. (1995). *Application of Remote Sensing and GIS to Change-Detection in the Rural-Urban Fringe Area of Bangkok, Thailand.* Thesis No. NR95-12. Space Technology Applications and Research Program, Asian Institute of Technology, Bangkok, Thailand.

Wilkinson, G.G. (2004). "A Review of Issues in the Intergration of GIS and Remote Sensing Data." *International Journal of Geographical Information Systems.* Vol. 10 No. 1, pg. 85-101.

World Commission on Environment and Development (WCED). (1987). *Our Common Future*. Oxford, England: Oxford University Press.

Xiuwan, Chen. (2002). "Using Remote Sensing and GIS to Analyze Land Cover Change and its Impacts on Regional Sustainable Development." *International Journal of Remote Sensing.* Vol. 23, No. 1, pg. 107-124.

Yeh, Anthony Gar-on and Xia Li. (2001). "Measurement and Monitoring of Urban Sprawl in a Rapidly Growing Region Using Entropy." *Photogrammetric Engineering and Remote Sensing*. Vol. 67, No. 1, pg. 83-90.

Yeh, Anthony Gar-on and Xia Li. (1998). "Sustainable Land Development Model for Rapid Growth Areas Using GIS." *International Journal of Geographical Information Science*. Vol. 12, No. 2, pg. 169-189.

Rozenstein, O. Karnieli, A. (2011) Comparison of methods for land-use classification incorporating remote sensing and GIS inputs. *Applied Geography* 31:2, pp 533-544 Online publication date: 1-Apr-2011.

Ruza, J., Kim, J., Leung, I., Coyan, C., Li, A., Kam, C., and Ng, S. (2012) Sustainable, Age-Friendly Cities: An Evaluative Case Study. ICSDEC 2012: pp. 196-203.

Tziztiki, J. G. M., Jean, F.M. and Everett, A. H. (2012): "Land cover mapping applications with MODIS: a literature review," *International Journal of Digital Earth*, vol. 5, no. 1, pp. 63–87

Yuan, Ding, Christopher D. Elvidge, and Ross S. Lunetta. (2003). "Survey of Multispectral Methods for Land Cover Change Analysis." Ross S. Lunetta and Christopher D.

Elvidge (Eds.). *Remote Sensing Change Detection: Environmental Monitoring Methods and Applications.* Ann Arbor, Michigan: Ann Arbor Press, Inc.

Zeppel, H. & Hall, C. (2002) *Arts and Heritage tourism.* In Weiler, B. & Hall, C. (eds.) Special Interest Tourism. London: Belhaven, pp.47-68

Zha, Daojiong. (2000). "Introduction: Globalization and Local Governments in East Asia." *East Asia: An International Quarterly.* Summer, 2000.

Zhang, Chunsun and Emmanuel Baltsavias. (2000). "Knowledge-Based Image Analysis for 3D Edge Extraction and Road Reconstruction." *International Archives of Photogrammetry and Remote Sensing (IAPRS)*. Vol. XXIII, Amsterdam.

Zhang, Chunsun, Emmanuel Baltsavias, and Armin Gruen. (2000). "Knowledge-Based Image Analysis for 3D Road Reconstruction." Paper presented to the ACRS 2000,

Taipei, Dec. 3-8th, 2000. *Proceedings of the 21st Asian Conference on Remote Sensing*, pg. 100-105.

Zhang, Yun. (1999). "Optimization of Building Detection in Satellite Imagery by Combining Multispectral Classification and Texture Filtering." *ISPRS Journal of Photogrammetry & Remote Sensing.* Vol. 54 (1999), pg. 50-60.

Web references

http://en.wikipedia.org/wiki/Remote_sensing

http://nexa.polito.it/

http://okfn.org//

http://www.aag.org/

http://www.asita.it

http://www.mapquest.com

http://www.esri.com and try the live demos

- sites of some major GIS software vendors

http://www.esri.com

http://fromgistors.blogspot.pt/?m=1

http://www.intergraph.com

http://www.autodesk.com

- some other introductions to GIS

USGS GIS Tutorial - http://www.usgs.gov/research/gis/title.html

The Geographer's Craft - http://www.utexas.edu/depts/grg/gcraft/notes/intro/intro.html

Nick Chrisman's "What is GIS?" - http://faculty.washington.edu/chrisman/G460/Lec02.html

ESRI's About GIS –

http://www.esri.com/library/gis/abtgis/what_gis.html

The Essential Guide to GIS - http://giswww.kingston.ac.uk/ESGUIDE/start.html

"What is GIS?" from Australia - http://www.dlsr.com.au/whatgis.htm

http://qgis.org/en/community/qgis-case-studies.html

http://www.gpsworld.com

www.ingramcontent.com/pod-product-compliance
Lightning Source LLC
LaVergne TN
LVHW072258300926
849160LV00018B/1693